RENEWALS 458-4574
DATE DUE

Women Farmers in Africa

WITHDRAWN
UTSA LIBRARIES

WOMEN FARMERS IN AFRICA
Rural Development in Mali and the Sahel

Edited by
LUCY E. CREEVEY

SYRACUSE UNIVERSITY PRESS 1986

Copyright © 1986 by Syracuse University Press
Syracuse, New York 13244-5160

ALL RIGHTS RESERVED

First Edition

Photographs by Michel Renaudeau appearing on pages 55, 61, 113, and 171 are reprinted with permission from *Au Coeur du Mali* by Michel Renaudeau (Paris: Editions Delroisse, no date).

The paper used in this publication meets the minimum requirements of American National Standard for Information Sciences—Permanence of Paper for Printed Library Materials, ANSI Z39.48-1984. ∞

Library of Congress Cataloging-in-Publication Data

Main entry under title:

Women farmers in Africa.

(Contemporary issues in the Middle East)
Papers presented at the Bamako Workshop on Training and Animation of Rural Women, sponsored by the Comité international de liaison du Corps pour l'alimentation (CILCA), held June 7–9, 1983 in Bamako, Mali.
Bibliography: p.
Includes index.
1. Women farmers—Africa—Congresses. 2. Women in rural development—Africa—Congresses. 3. Women in agriculture—Africa—Congresses. I. Creevey, Lucy E. II. Bamako Workshop on Training and Animation of Rural Women (1983) III. Liaison Committee for Food Corps Programmes International. IV. Series.
HD6073.A292A358 1986   331.4'83'096   85-27771
ISBN 0-8156-2358-5 (alk. paper)
ISBN 0-8156-2359-3 (pbk. : alk. paper)

Manufactured in the United States of America

To my husband, Ton, and my sons, Rob and Ari,
who cheerfully tolerated my preparation of the Bamako Workshop
and this book, and to the women of UNFM
and of the Malian rural development agencies whose work
provided the inspiration for this book.

LUCY E. CREEVEY is Director of the Program in Appropriate Technology and Energy Management for Development and Professor of City and Regional Planning at the University of Pennsylvania. Her interest in the Sahel began in 1960 when she first visited Senegal. Since then she has frequently traveled to Senegal and other Sahelian countries for research and as part of her work as consultant planner to CILCA, a nonprofit program sponsoring projects to increase food production at the village level. She is the author of *Muslim Brotherhoods and Politics in Senegal*. She received her Ph.D. in political science and African studies from Boston University.

Contents

	Foreword	ix
	Preface	xv
	Introduction	1
Section I	**Women Farmers in Mali and the Sahel**	
	Commentary	16
1	Sex Roles in Food Production and Distribution Systems in the Sahel *Kathleen Cloud*	19
2	The Role of Women in Malian Agriculture *Lucy E. Creevey*	51
3	The Role of Women in Rural Development in the Segou Region of Mali *Mariam Thiam*	67
4	The Changing Role of Women in Sahelian Agriculture *Bernhard Venema*	81
Section II	**Development Programs for Rural Women in Mali and the Sahel**	
	Commentary	96

5	The Ouélessébougou Training Center for Rural Women Extension Agents *Halimatou Traore*	105
6	Training of Rural Women With the Stock-Farming Development Project in Western Sahel *Coulibaly Emilie Kantara*	117
7	The Grassroots Women's Committee as a Development Strategy in an Upper Volta Village *Helen Henderson*	133
8	Activities of the Women's Promotion Division of the National Functional Literacy and Applied Linguistics Board (DNAFLA-DPF) *Dembele Sata Djire*	153
9	Improving Women's Rural Production Through the Organization of Cooperatives *Sacko Coumbo Diallo*	159
10	Appropriate Technologies for Women of the Sahel *Jacqueline Ki-Zerbo*	167
11	The Lorena Cookstove: Solution to the Firewood Crisis *Jonathan B. Tucker*	179
	Conclusion	189
	Appendix	197
	Bibliography	199
	Index	209

Foreword

Food Corps Programs International (CILCA) came to Bamako in June 1983 to learn more about the efforts of African women farmers in Malian village projects, such as those in CILCA's Toko (Segou) and Katibougou projects. At first, CILCA organizers thought the Bamako workshop should deal specifically with the constraints on women farmers in these Sahelian projects. The group sponsored another workshop, organized by Zimbabwean colleagues in Harare (See CILCA 1983b). It soon became clear, however, that more than operationally specific information was needed. We had to make an intellectual effort to challenge prevailing development paradigms as well as practices, and therefore reorganized the workshop to draw together scholars and practitioners.

The workshop, cosponsored by the Union of Malian Women, attracted broad interest. The Mali government sent participants from various ministries involved in rural development, and some hundred people attended. The papers were lively. The discussion was frank, often heated, and many significant topics were debated (see CILCA 1983a). It is still going on, and this book, edited by Dr. Lucy Creevey, is an outcome of continuing reflection.

Since 1980, CILCA, an international liaison committee among Third World practitioners in rural development, has been seeking faster progress for the rural poor. We try to apply lessons from successful projects around the world that can yield principles to guide each national Food Corps Program. In specific regions of Mali, Tanzania, and Zimbabwe, together with experienced colleagues from Sri Lanka and Mexico, national CILCA teams are implementing integrated rural development projects.

CILCA's goal is the reduction of rural hunger and poverty through people's participation in improved agricultural planning and production. CILCA's methods are based on certain assumptions about appropriate development methods, development as a field of study, and the relationship between society and production.

We know poor rural communities remain limited in their ability to sustain progress as long as production stagnates. Increased production requires better farming techniques. However, CILCA is not just interested in increasing agricultural production, but also in strengthening the entitlements of all—women, children, and men—to better nutrition, health, housing, and education. Attaining these goals requires emphasis on method and on communication with producers.

CILCA's method combines research in farmers' communities with practice. Data collection is important; we must know about the farming community, its institutions, technology, productivity, markets, household consumption, land tenure, division of labor, entitlements, and access to agricultural inputs and services. Such data is vital to project preparation, monitoring, and evaluation. However, more than data collection is needed to be effective in development; it requires understanding of basic social processes and structures.

Since Africa is a latecomer to development, much of the prevailing wisdom is rooted elsewhere in the Third World and has little basis in African reality. In Africa, soil conditions, rainfall, social organization of production, marketing, and access to agricultural institutions vary greatly within villages and among regions. Unlike Asia, Africa has mostly rainfed agriculture. Soil is more fragile and less fertile, population density is lower, transportation costs are higher, and the social organization of agricultural production is quite distinctive.

The Green Revolution has scarcely touched Sahelian villages. Low income Malian farming families have to survive under some of the harshest environmental conditions in the world. "The husband is customarily supposed to supply only a house and grain to his wife in Bambara society; the rest is the responsibility of the wife," writes Mariam Thiam in her essay here. The wife contributes to the livelihood of the household by farming or trading to obtain vegetables, fish, chicken, and condiments. Sahelian village women try to earn enough to buy cloth and other necessities for themselves and their offspring. Society expects village women to take care of their own fields and accomplish certain set tasks in the men's fields. At the same time, they have the responsibility of gathering firewood, carrying water, cooking, cleaning, nursing, and caring for children. The woman's working day is therefore longer than that of the man.

On top of unequal customary division of labor between men and women came further inequities after the introduction of cash crops. The colonial rulers imposed inflexible taxes. They went up after independence, and villagers must pay what seems an endless round of national and regional taxes, as well as dues to parties and women's and youth associations. As poor families find it essential to have an outside income, many men leave the village to look for paid work; then the women are left to perform all the farming tasks.

Labor and poverty are growing constraints in Sahelian agriculture. People have no money for costly inputs or tools; per capita income was $180 in 1982. Though the population is growing, the percentage of labor in Mali's agriculture declined to 73 percent in 1980 as compared to 94 percent in 1960. In the early 1980s, Mali's index of food production per capita was only 83 (1969–1971 = 100) (World Bank 1984, 218, 228 258). There is an urgent need to find strategies to reverse the agricultural decline—strategies that must rest, in part, on social analysis.

Looking at women farmers more closely becomes a necessity. We know women carry a disproportionate share of village work involved in food production. We know that CILCA projects, in order to come up with ways to reenforce village institutions so as to favor greater production of food both for consumption and for the domestic market, need better communication with women farmers. CILCA's national technical teams include more men than women, a reflection of the realities underlying recruitment for scarce places in African institutions of higher learning. In each African project area, male technicians face social constraints governing contact between the sexes. In Toko village, where a woman was part of the technical team, it was possible to make quick contact with the rural producers in the community. There, the women, partners in the struggle to wrest a better livelihood from the soil, responded quickly to new opportunities (Morgenthau 1985).

The social and economic roles of women and men in different communities are related but distinct. So are some of their institutions. What is the impact of any projected changes on the customary division of labor between men and women? On the customary return for labor? Each producer's will and ability to participate in development activities are related to answers to such questions. Each producer's interests must be considered if real incentives and realistic avenues to productivity are to be identified.

Some crops in some areas, vegetables for example, are mostly women's crops. If a plan calls for more of women's crops, what is the effect on men's crops? Other crops, such as cereals in some areas, are mostly men's crops. What will the effect of increasing men's crops be on the burden of labor on women and on their abilities to keep producing women's crops? There is a division of labor by sex in animal husbandry also. Without attention to these divisions that reflect a balance among men, women, and children in household and community, technicians can get unexpected counterproductive results. Producing more of one crop might rest on impeccable economic logic, but it could damage the household and community. For example, cash crop production can go up while nutrition and child health go down, and more people become vulnerable to hunger. This realization raises more questions.

How can we plan effectively for integrated development involving women as well as men? How can we be sure women get resources to produce their crops? How can we reach women farmers and train them? How important is it to have women technicians? Should there be projects and services for women only? How can we reduce household claims on women's time, and free time so women can produce more?

CILCA's Bamako workshop turned into a wide ranging discussion of important questions that can help in the design, implementation, and evaluation of village development projects. The workshop made us more sensitive to the interaction between concrete social reality and agricultural techniques. Each low-income village and region faces technical problems, and some technical answers already exist, at least in part. Mexican experience with growing corn in traditional farming areas can be useful in corn growing areas of Tanzania or Zimbabwe. Sri Lankan experience with the *ipil-ipil* tree *(Lucaena)* might help West Africans seeking fast-growing, deep-rooting, drought-resistant trees. Research in farmers' fields can yield many answers to questions about water retention, fertilizing, crop varieties, and cultivation techniques. Of course, technical questions remain; sorghum, millet and cassava have yet to yield their secrets to scientists. But Africa's agricultural problems, we know, are only in part technical. The institutions hold many secrets about production, and these require quite different forms of investigation than those of the laboratory.

Those of us who took part in the workshop in Bamako came away convinced that rural Sahelian women are the "secret agents of modernity." Women have a big stake in the survival of all members of the household. They yearn for better results to reward their endless labors. Hunger and poverty is hardest on them. Women have the best reasons to become fully engaged in participatory rural development action. Sahelian traditions point them in that direction.

Technical and development planners must communicate with them. Once African women farmers find ways to make their labors more productive and turn their institutions towards development, there will be no stopping them. They want to help themselves and, given support, can show their governments how to make national agricultural policy more effective.

CILCA's energetic Executive Secretary, Aly Cisse, coordinated the workshop. The Canadian International Development Agency and the Malian government generously helped underwrite the costs. Many Malian women taught the workshop organizers how to think about the problems. Practitioners and students of rural development are in their debt.

March 29, 1985
Ruth S. Morgenthau
President, CILCA

Preface

This book contains a core of papers which were presented at the Bamako Workshop on Training and Animation of Rural Women sponsored by CILCA in 1983. It is thus a by-product of CILCA activities, although the board of CILCA is in no way responsible for the interpretations and conclusions of the editor or of the individual contributors.

Credit must be given to Vera Gravier in the preparation of this book. Mme. Gravier was the official interpreter at the Bamako conference. She not only translated from French to English and from English to French, but also helped some participants write their presentations, wrote a summary report of the proceedings, and translated all the papers into English at the conclusion of the Workshop. Her translations, which were approved by the CILCA staff in Mali and the Union Nationale des Femmes du Mali, are accurate renditions of what was presented in French. They are the basic translations used in this book for the papers by Mesdames Traore, Diallo, Kantara, Djire, and Ki-Zerbo. However, some modifications have been made in the interest of clarity. The papers were written for oral presentation. They typically had large numbers of lists, many short sentences, and short elliptical references which, when presented orally in the context of the Workshop, were clearly understood. In order that these papers be intelligible for readers not directly familiar with their work, I have taken the liberty of completing sentences, making lists into paragraphs, and clarifying references by adding a word or two when necessary. I have done this (based on the French texts and my notes from the Workshop) without changing the original meaning. The original, unedited version of these papers may be found in the proceedings of the Bamako Workshop distributed in French by CILCA.

Finally, I am indebted to many people for their help in the preparation of this book. In particular, I want to thank Ruth S. Morgenthau who encouraged me in the first place, Margaret Stephens for her invaluable help in polishing the rough manuscript, and Bobby Russo for her long, painstaking work in typing and preparing the full manuscript.

LEC

May 1985

Women Farmers in Africa

Contemporary Issues in the Middle East

Introduction

There is a pervasive bias . . . against the technology and needs of rural women. Until recently, little attention was paid to home gardens and backyard farming, often sources of small but vital incomes for women. Domestic technology—for processing food, cooking, cleaning, sewing, fetching firewood, carrying water—all traditional responsibilities of rural women, is regarded as uninteresting, a low priority. When the person-hours devoted to these activities are considered, and the drudgery they entail, it is a grave reflection on those with power how miniscule has been the attempt to improve the technology of such activities. The processing of staple foods (cassava, millets, sorghum, paddy) by hand is a grueling task for . . . millions of women . . . yet few engineers and scientists have turned their minds . . . to seeing how the process could be made easier.

The pro-male and anti-female bias applies in other spheres too. Ploughing, mainly carried out by men, has received more attention than weeding or transplanting, mainly carried out by women. Cash crops, from which male heads of household benefit disproportionately, have received more research attention than subsistence crops, which are more the concern of women. Even now, after a massive shift of rhetoric . . . not much more than a modest foothold has been established. . . . It is rare indeed to find substantial changes in perception, attitude or behaviour among the male majority of professionals. Scientific and engineering establishments in particular remain heavily male-dominated and are usually still a very long way from recognizing, let alone giving attention to, the needs of rural women. (Chambers 1983, 80–81)

Background

There is increasing concern among those interested in the development of rural areas in the Third World about the lack of attention paid to improving the lives and productivity of rural women. This concern usually is expressed in terms of two related themes. The first, and most common, theme describes the injustices suffered by rural women, the inequality to which they are subject, and their lack of political and social power resulting in poorer living conditions and harder work for them than for their male counterparts. The overall social cost of this unjust treatment of over half of the adult rural population is often pointed out. The statement by Robert Chambers, quoted above, expresses some of this concern.

The second line of discussion relates to practical questions about how the productivity and income-generating capability of the rural population as a whole may be increased. When the significant role of women in cash crop[1] and food crop production and animal husbandry is understood, the lack of programs to improve their agricultural knowledge and increase their productivity appears suicidal. Rural development will not take place unless a major effort is made to work with rural women. Perhaps in the distant future the roles of men and women will be differentiated as in some of the more industrialized countries, and women will not be responsible for the heavy physical labor of farming. But, if that does occur, it will not happen for a long time. In the meantime, poor countries can not afford to continue to focus major aid and training programs exclusively on men.

This book relates most closely to the second theme—the impact of efforts to work with rural women. It follows the publication of many books, articles, and monographs which discuss the situation of women in the Third World (see Boserup 1970, 1980; Presvelou and Spijkers-Zwart 1980; Agency for International Development 1978, 1982; Obbo 1980; Nelson 1981; Lewis 1982). Many of these studies have been of considerable assistance in clarifying the actual role of women in the social and economic systems of rural areas. Beginning with the ground-breaking work of Esther Boserup (1970, 1980), numerous scholars have given painstaking attention to documenting what women do, what the signifi-

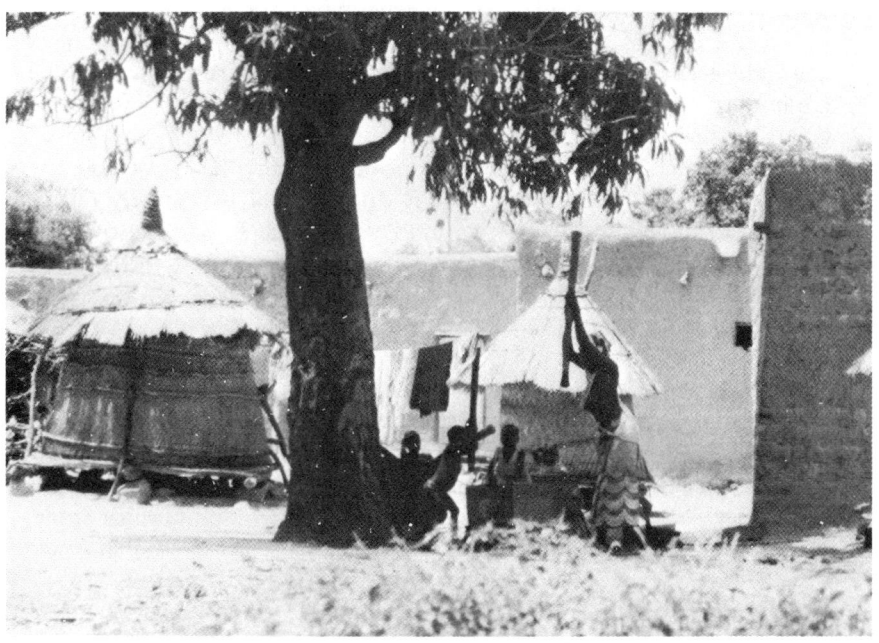

This Bambara woman in Mali is grinding her grain in the traditional manner outside the hut where she and her children live within her husband's compound. Photo by Richard Harley

cance of this is to them individually and to society as a whole, and what has been done to assist them (or what has not been done) and why. But, while these many studies have made clear the perilous situation of rural women and the need—in general terms—to do something immediately, they generally have not gone into depth on what actually must be the substance of rural women's programs and what problems arise when work is begun among and for rural women. In the 1980s, there is widespread acknowledgment of the failure of many programs to achieve at least their initial objectives. Without carefully examining what has been done, and drawing lessons from this experience, future programs will continue to make slow progress. Thus, although this book touches on some of the same issues as did earlier studies, it differs from most in its specific focus on planning problems relating to working with women in the agricultural sector. The sole purpose of this book is to provide direct cases of ongoing projects together with specific analyses of the issues which these experiences raise. As such, it may be of interest to two sets of people: those who want to know the real issues faced when ideas and theories of rural development are implemented and those who have a particular concern for how programs relating to rural women actually work.

The papers included here make principal reference to one country, Mali (West Africa), although there is discussion of two closely related Sahelian examples as well: Senegal and Upper Volta (Burkina Faso).[2] The decision to focus on one area, rather than include examples of projects throughout Africa, was in the interest of depth and clarity. By looking at Mali specifically, with more general reference to other Sahelian cases, it is hoped the specialist with no special expertise on Africa will be better able to understand what the conditions of life are, what resources are available, what has been done, and the impact of what has been done. It should be possible, therefore, to draw conclusions about the utility of some of the assumptions made and strategies followed in rural women's projects. The Sahel, (see Appendix) furthermore, is a particularly good case for study, being one of the poorest regions in the world with its agriculture in a state of severe crisis.

To directly address the issues which trouble those actually responsible for programs in the field, this book includes a majority of papers written by Malian women who direct the programs for rural women in that country. These are not academics who sit back and abstractly theorize on how best to handle a particular problem or who insightfully point out general flaws in projects completed elsewhere. These women do not have the time to do a literature search for what other people have said or done about women in agriculture. As a result, their papers lack the kind of sophisticated theoretical arguments which customarily are found in published scholarly papers. But these are women who know from firsthand experience what the problems are from day to day. They know what works and what does not work in their familiar context. When they write, they are writing about what is happening now with programs aimed to promote rural women in their country. As such, their comments are more valuable for those who want to understand rural development than the more polished presentations of the "experts." A few papers written by "experts," however, are included to round out the discussion. These contributors are closely familiar with the situation of women farmers in the Sahel and Mali, and with projects that have been introduced among rural women in those areas.

The inspiration for this study was a Workshop held from June 7–9, 1983 in Bamako, Mali, sponsored by the Comité International de Liaison du Corps pour l'Alimentation (CILCA),[3] funded by the Canadian International Development Agency (CIDA) and organized by CILCA staff together with the Malian National Women's Association, the (Union Nationale des Femmes du Mali—UNFM). Originally, the Workshop was planned as an exchange of ideas among 30 people, including 10 foreign experts on the subject of women in rural areas and approximately 20 Malians (CILCA-Mali project managers, UNFM members and government representatives). The purpose of the Workshop was to highlight the important factors to consider in the training and mobilization/*animation*[4] of rural women as a backdrop to detailed workplans being prepared for two CILCA-Mali projects. In fact, the Workshop turned out quite differently because of the interest and concern of people in Mali. The future CILCA projects were not the focus of interest. More than 100 people attended the meetings to listen to the papers which dealt with actual projects in Mali and to discuss problems and prospects in general. Three foreign experts presented papers—a Dutchman, an American woman, and an Upper Voltan woman—and two papers were presented on existing CILCA-Mali projects. The papers which dominated the sessions were those by representatives from UNFM and the Malian government on work in progress. The discussions which followed were lively and there was often substantial disagreement over points which were raised.

These papers were impressive in that they showed the very wide range of programs begun among rural women in Mali. They indicated a substantial depth of understanding, not only of the problems of rural women and the causes of these, but also of the most current strategies for confronting these problems. But these papers, and the debates which followed, also graphically illustrate how much is unresolved in the whole field of planning for and with rural women, and how much of a need there still is for discussion of the issues involved.

History of Rural Development

It is not only in regard to working with women that there are unresolved issues in rural development planning. In fact, the whole history of rural development planning is one of controversy and indecision. Many false assumptions made in the past about economic and social development haunt planners today, either because they persist or because the results of these beliefs are difficult to overcome. Most planners now assume that development is the process of change which brings about improvements in the standard of living for all the population (i.e. not benefitting only one sector without improving the lives of most others). Unfortunately, general acceptance of this premise is relatively recent. In the 1950s and 1960s, economists still discussed unequal development assuming that it would be wise to invest all funds, time, and energy into industrial development even if this left agriculture stagnating or actually decreasing in productivity. Unequal development, it was asserted, would eventually improve everyone's life, even in the "peripheral" sectors where there was no investment.[5]

Where agriculture did receive attention, it was largely in regard to producing cash crops to increase the flow of capital from abroad. It was assumed that productivity would increase most rapidly with large plantation-type projects which used sophisticated agricultural technology. Thus, both machinery and the accompanying technical advisers were imported from the outside. The small individual farmer who practiced a simpler traditional agriculture was largely ignored or included as unskilled labor for the large agricultural projects. Women farmers as such were ignored altogether.

Inevitably, the flaw in this approach emerged. Plantation agriculture worked as long as the outsiders continued to invest in it and to provide assistance. But productivity was never quite what had been projected based on research at national agricultural centers. Machines constantly broke down and spare parts had to be imported. Workers often did not stay long enough to be properly trained in the new techniques before they had to be replaced and the training begun all over again. And the new approaches did not spread widely. Although local farmers often began to grow new crops for which they could get cash, they did not use the methods of the plantations. Thus, for example, local cocoa farmers produced cocoa beans inferior to those on French-owned plantations during the immediate post-World War II period in the Ivory Coast (see Morgenthau 1964, 167–174).

By the mid-1970s, evidence suggested that lack of attention to agriculture had resulted in slowing down the economic growth of developing nations. The benefits from industrial growth had not quickly trickled down in appreciable amounts to the rural sectors. Instead the gap between rural and urban regions had widened. The new services—health, education, and welfare—were located in or near cities where virtually all industrial investment took place. Wage paying jobs were available only in the cities. As a consequence, migration from the rural areas had swelled, depleting the active work force left for agriculture. Governments, faced with disgruntled urban populations unable to find jobs, had attempted to maintain calm by providing food at a price which the urban dwellers could pay, thus controlling the sale price of food crops and further reducing the incentives for the development of the agricultural sector. The result was an increasing dependency on imported food.[6]

In an effort to redress the internal balance and to promote more national self-sufficiency, donor agencies and governments of the Third World finally focused their attention on the rural sector. Thus, the World Bank staff in the mid-1970s wrote with pride of being the "largest single external source of funds for . . . agricultural development" as a result of their changed perception of development and the importance of the rural sector to that process (World Bank 1976). All donor agencies began to talk about equalizing the benefits of the development process and about reaching the "poorest of the poor." The poor were disproportionately located in rural areas—85 percent, according to the World Bank (1976, 4). The concept of first meeting, "basic human needs" for all people, was advanced as a strategy and/or criterion for project selection (see Streeten 1981). To provide these basic human needs, projects were to be directed toward all types of agriculture, animal husbandry, poultry farming, small scale industry projects, and the provision of infrastructure, education, health, and welfare services for the neglected areas.[7]

Coincident with the new emphasis on rural development was a new international emphasis on food production. Although malnutrition, hunger, and actual starvation were not new problems by any means, in the 1970s, politicians began to rediscover them. This rediscovery was not purely political however. It reflected increasing pressures, evident in the 1970s, on food supplies in the developing areas. In the Sahel, for example, disastrous droughts underscored the lack of sufficient food for large numbers of people in the region. The World Food Crisis was now widely spoken of and discussions began as to what emergency and long term measures might be applied (for example at the World Food Conference in Rome in 1974).

As a result, rural development projects began to emphasize support of food crop production rather than cash crops only, and to regulate credit, marketing, and pricing policies to encourage—even enable—farmers to grow more and better food. Although there was some disagreement as to the efficiency of "bottom-up" as opposed to "top-down" planning programs (see Stohr and Taylor 1981), attention shifted to the individual producer, the small farmer rather than the manager of the plantation-type system. Small subsistence farmers, of whom a great many were women, had provided the bulk of food crops in the pre-colonial era. Moreover, they were a large proportion of the rural poor. For a country to become self-reliant in food production, and certainly for the basic needs of the rural poor to be satisfied, those same small farmers had to be reached with new programs. These new programs needed to assist small farmers to obtain tools and other inputs, training, credit, and marketing facilities, all of which they would be responsible for themselves. Without such personal involvement by local individuals, food crop production projects would fail just as so many other agricultural endeavors had failed in the past (see Lofschie and Cummins 1982).

Once the need to work with local farmers was accepted, programs had to be developed which would successfully lead to a permanent increase in productivity and standard of living. Considerable uncertainty remained as to which institutions or programs would be most effective in assisting the new development efforts.[8] Planners now realized that, just as concentrating on one sector of the country could not bring overall development, so concentration on one dimensional projects could not result in improvement in all facets of life and might even mean the failure of the activity chosen for attention. If the economy was to develop, so must the health and welfare of the people. Development implied an interdependent system of change and growth. Thus, much lip service was given to so-called integrated rural development planning. In principle, this meant that projects which simply tried to increase or improve one activity, such as the production of a given crop or the establishment of an income generating activity, would no longer be established. Rather, planners would consider other facets of life and, ideally, plan for improvements in all related areas as well. All of village life would be simultaneously affected—wells dug, quick-growing firewood trees planted, smokeless stoves introduced, grinding mills for domestic grain preparation made available, health and education programs begun, roads improved, credit and marketing facilities established, training for rural industries provided, vegetable crops introduced, and, finally, the production of food grains and cash crops better balanced and the crops themselves improved (with better fertilizers, selected seeds, tools, and training). A program which did all these things, it was thought, had a chance of really transforming rural society and making the work of both men and women easier and more productive.

The trouble was that it was difficult, if not impossible, to do everything at once, even with unlimited resources. The rural village is a bastion of tradition established through decades of experience. Changes occur over the years in traditional society but they are usually gradual (unless in response to a disaster). Villagers cannot be expected to understand and carry out extensive reforms simultaneously in all that they do. Given sufficient resources, they could be forced to adopt new ways of doing everything, but these imposed changes would not be fully internalized or permanently rooted. Besides, resources are severely limited in most cases. Except for a few token villages—the home village of the president, for example—the government of a developing country cannot afford to provide all the support needed for a full social and economic reform of village life.

Lack of money for tools, credit, infrastructure, etc. is not the only bottleneck—the scarcity of extension officers is at least as significant. There are not enough trained government workers and project officers to assist villagers to understand new techniques and new approaches to their daily activities. This type of assistance, both for learning technical skills and general mobilization to enable villagers to choose programs and execute them themselves, is the key to the success or failure of most programs. But such training can only be done over several years. Even if villagers are initially very enthusiastic about a proposed program, they will need backup assistance for a long period of time to help them adapt to the changing situation which new crops, new health and education programs, new tools, etc. help create, and to assess what modifications need to be introduced given the way things develop. It is too easy to give up when an unpredicted drought or flood occurs, when a machine (even a simple one) breaks down, when a disease or a plague of locusts hits a particular crop. There are many examples of failed projects to illustrate this point—badly needed biogas generators left unmaintained and gradually becoming unusable, wells in regions where water otherwise had to be hauled by hand over many miles which have nonetheless not been cleaned and have silted in, crop programs where farmers, after the first few years, have gradually gone back to their old production methods and lower yields.

The villagers have many explanations for these lapses—they did not understand what to do to the generators, the wells filled in faster than they could clean them, the new crop methods were not as good as the old. Government agents and foreign planners have their explanations as well, which all too frequently have to do with the stupidity and laziness of the villagers once no one is controlling them and forcing them to conform to the project. But, in reality, the villagers did not understand the proposed changes to the point where the altered behavior could become the accepted way of doing things. They needed a much longer exposure to the benefits from altered behavior and to the results of the different things that they did. One extension agent, responsible for sixty villages and forty thousand people, could never provide this kind of support.

Sometimes, perhaps even frequently, the fault lies in what is proposed as a projected set of reforms. All too often, planners wrongly believe that they know better than the villagers how things should be done. But subsistence agriculture is a survival response to very difficult living conditions. Farmers know what the soil will do, what the other climatic factors are, and what their choices are (in the absence of new technologies and tools, improved seeds, and chemical fertilizers). Once something new is introduced the whole system changes, sometimes with unexpected negative results. There are too many instances of unfortunate results of such "reforms." The smokeless stove, for example, which had proved quite successful in South Asia, did not spread as quickly or easily in parts of West Africa. The various social functions of the traditional stove had not been understood, and had not been provided for by the proponents of the new type. Agronomists in Mexico who had pushed enthusiastically for growing corn without intercropping pole beans had to drastically revise their program when they found that the traditional intercropping might produce less corn, but in fact produced more food and a better balance in terms of what local people wanted to eat.

Programs for Rural Women

In this context, it is not surprising that many programs for rural women failed when they were first introduced, or at least did not achieve the goals which had been set for them. There was little understanding of the actual role women played in the rural economy and even less knowledge of successful ways of supporting their activities. Moreover, men, often in high government positions, felt that, given scarce resources for any development work, investing in programs for women was a luxury which they could not afford. Women, it was argued, would benefit from the programs directed to rural men who were the household leaders and the ones whose opinions and behavior had to be changed first. Or, unwilling to sound uninterested in projects for rural women so popular with foreign donors, these government leaders would pay lip service to women's programs but would in fact give them the lowest priority in the whole scheme of things they set out to do. Thus, little money would actually go into the women's programs and the best staff would be allocated to other work. Automatically, women were given charge of women's programs, whether or not they were properly trained or motivated for this work.

Gradually over the last ten years, governments of developing countries have given more serious attention to women's programs. This is due to two factors. In the first place, improving agricultural production and rural development in general proved to be even more difficult than originally anticipated. It was not just a matter of allocating sufficient resources, but, as the earlier discussion above indicates, of knowing what to do with them. Secondly, women of developing nations have become more and more vocal in expressing the needs of rural women and the importance of working with them. It was increasingly difficult to ignore their arguments as more and more evidence on the ineffectiveness of rural planning to date began to be publicized.[9]

Programs for women still have second-class status in rural development planning in most Third World countries. But more than ever before, work is being done, and a track record is gradually emerging on what is successful, what is not, and what we still do not know. At least one point has been clearly understood and accepted. Just as establishing a project among sedentary farmers will not necessarily help neighboring pastoralists, organizing and equipping male farmers will not automatically provide assistance to the work done by women farmers. It is impossible to work with women in isolation; the awareness and consent of their male family members is essential to success. Nonetheless, the focus must be directly on women—their special needs, their particular capabilities, and their state of training must all be understood and utilized in plans which aim to improve their lives and their productivity. Moreover, all these factors differ depending on the ethnic background, geographic location, religion, social strata, and economic conditions. Special techniques for motivating and training rural women and for capacitating them to undertake their own development must be devised. The methods used for rural men may or may not be appropriate, depending on the women involved and the specific ways they are to be helped.

Because serious efforts to work with rural women in the developing world are so new, there is still a great deal to learn about what will succeed in a program or project. This book attempts to contribute to the general understanding of what is involved in setting up programs and what have been major hindrances to their success. The second section, Development Programs for Rural Women in Mali and the Sahel, is a series of papers which discuss a variety of projects in Mali and Upper Volta. These papers have been chosen to illustrate both the range of programs for rural women and some of the major issues and/or problems which arise in carrying them out. In many cases, the papers discuss fundamental assumptions about what will bring about change. In others, the assumptions are implicit in the project arrangements which are presented. Some of the more interesting issues which various contributors bring up were not actually perceived as issues by the author of the paper. By contrasting the papers with each other, clear differences of approach can be seen. These differences were emphasized in the debates among the women at the Bamako workshop which followed the original presentations. Since the debates were wide ranging and unsuitable for inclusion here, the Commentary by the Editor and the Editor's Notes (which precede each paper) attempt to describe the major points of disagreement and the major problems which the papers bring out. The original presentations, however, merit close reading as most of them are straightforward, succinct accounts with no theoretical padding. They are thus the best possible sources for understanding the commitment of those in charge of women's programs, what they do and how they do it.

The first section of the book, Women Farmers in Mali and the Sahel, includes four papers which discuss what women do in the rural sector in Mali and the Sahel. While it might seem to the uninitiated that a general introduction to what women do would suffice, there is much that is still unclear on the subject. The work of Esther Boserup (1970) made a tremendous contribution to the study of rural women, but some of the generalizations she presents, which have been accepted as gospel for all of Africa in many later studies, may well be inappropriate in Mali and similar countries. This is a serious issue, for assumptions resulting from her work—such as the belief that women do not grow cash crops—affect the choice of development strategy. Again, the Commentary and Editor's Notes which precede the papers in this section attempt to draw out the differences among them and the major points which the papers present. The papers, of course, present a far more in-depth discussion of the subject than the brief Commentary or Notes can do.

Symbolically, there is no unified set of recommendations in the conclusion to this book. The issues and problems presented here are summarized as a step in the process of understanding how rural development works and what needs to be done. But even the contributors to this book did not have a common view of how to confront the problems they faced, although there was general agreement on some principles which had to guide the organization of projects. The editor, acquainted with numerous development studies in which the author has presented "the solution" (one which shortly after publication is rejected even by the author himself)[10], is reluctant to make any overall prognostication or set of guidelines. When there is no real answer, these efforts are at best embarrassing and at worst harmful, for they may lead to misguided actions in situations where resources are scarce and policy mistakes tragic. It is not that inactivity is recommended here: far to the contrary. But each situation, and all its resources and constraints, has to be carefully and individually evaluated. The program of action developed must be tailored to the needs, values and capabilities of the country (and/or region) where it will be carried out. What is most needed as a resource for this work is hard data on what really has happened when various strategies were followed. This book is a modest attempt to provide some of this information in regard to working with women farmers in the Sahel.

Notes

1. The term "cash crop" is most often used to refer to crops grown for export. Often these crops replace food crops grown for domestic consumption. There are, of course, food crops such as peanuts which are also export crops. The normal contrast, however, is between subsistence traditional agriculture grown for the needs of the rural population and commercial agriculture aiming at external markets (see Johnston and Clark 1982, 254).
2. The name of Upper Volta was changed to Burkina Faso in August–September 1984. Because the papers included here were written before that time, and for consistency, this book retains the name Upper Volta.
3. The Comité International de Liaison du Corps pour l'Alimentation (CILCA) is a private, non-profit organization which promotes projects in the Third World aiming to increase food production at the village level (see CILCA 1984).
4. There is no satisfactory English word for *animation*. It means mobilization with the connotation of awakening or inspiring. This book uses the term *animation* and mobilization and retains the French *animateur* or *animatrice* to describe the agent of this process.
5. The standard proponent of unequal development was Albert V. Hirschman (1968). Arguments for industrialization as a motor of development followed the neoclassical economics reasoning that capital accumulation was the major (sole) factor in development (see Nurkse 1953, Lewis 1955, Sweeten 1979).
6. Discussions of the results of "unequal" or "uneven" development policies with reference to the neglect of the agricultural sector may be found in virtually all current discussions of development planning (see, for example, Amin 1980, Friedman and Weaver 1978, Chambers 1983, Rothko Chapel Colloquium 1979, Bryant and White 1982).
7. Discussions of new theories of development with especial emphasis on rural development planning may be found in numerous sources (Chambers 1983, Ghai et al. 1979, Cheema and Rondinelli 1983, Rondinelli 1983, Griffin 1976, Johnston and Clark 1982, Leonard and Marshall 1982).

8. See, for example, the discussions of local institutions and the conditions under which they assist development efforts (Esman and Uphoff 1984, Esman and Uphoff 1974, Leonard and Marshall 1982, Ghai et al. 1979).

9. See critical analysis of World Bank projects in Africa in Lele 1975.

10. See Albert Hirschman's brief dismissal of some of his earlier premises (Rothko Colloquium 1979, Foreword, xv–xviii, and John Friedmann's similar disclaimer, Friedmann and Weaver 1979, pp. 126, 186–207).

SECTION I

Women Farmers in Mali and the Sahel

Commentary

Each of the four papers included here sheds a slightly different light on the questions surrounding the role played by women in Malian agriculture. The first paper is written by Dr. Kathleen Cloud, who takes a systematic look at the range and variety of tasks performed by rural women in the Sahel. She also presents the major factors affecting Sahelian agriculture and food production and the changing roles of men, as well as women, in Sahelian agriculture. She uses evidence from a United Nations Economic Commission for Africa report which shows that in all of Africa, in the traditional system, women perform 70 percent of the work in food production, 50 percent in storage, 100 percent in processing, 50 percent in animal husbandry, 90 percent in water supply, 80 percent in fuel supply, and 60 percent in marketing.

Dr. Cloud believes that women are losing power relative to men, and consequently, their independence in the process of modernization. Programs to promote rural development, she points out, frequently undercut the position of rural women. She ties this pattern of progressive undermining of women's roles in the Sahel to that observed by Esther Boserup (1970) elsewhere in Africa. Organizers of large projects, she points out, assume that men make all decisions and control all resources and therefore direct all assistance activities to them. The specific tasks of women, such as providing vegetables, fruit, milk, cheese, etc. are often ignored in aid programs which tend to stress cash crops. In particular, women are being closed off from access to resources they traditionally had, especially land which, as it passes into private ownership, is being registered officially only to men. Dr. Cloud writes: "Whatever the causes, the pattern of exclusion of women's productive activities from access to development resources plagues many of the current development projects in the Sahel (p. 44).

The second paper in this section is written by the editor. It is included to give background to further discussions about women in Malian agriculture. The paper attempts to present basic facts about agricultural conditions in Mali and to demonstrate the interdependence of men and women in both food and cash crop production. Thus, rather than seeing men replacing women in agricultural tasks as cash crops become important (as many followers of Boserup's theories have done), this paper shows that women in Mali grow cash crops in their own fields and work in the cash crop fields controlled by men as well. Furthermore, this paper contends, the impact of the modernization process on women in agriculture is best understood if the long, constant process of change in men and women's agricultural roles over time in response to changing circumstances in Mali is understood. Modernization programs did not simply alter a rigid, fixed power relationship; the situation was far more fluid and complex. The paper also demonstrates that there is not one uniform traditional agricultural role into which all women fit. Not only ethnic background, but also factors such as age, religion, status, position of the family, wealth in the family, etc. influence what women do.

This chapter does not minimize the extent to which women are, and have been, important in agriculture, or the severe problems which they now face, but it does stress the need for a consideration of the entire village/family agricultural system when devising programs to work with rural women.

Mariam Thiam's paper gives relatively greater emphasis to the plight of women in Mali as a result of the process of modernization. She concentrates on one region of Mali: Segou. She gives a clear description of the broad range of arduous tasks performed by women in this region, from gathering wood to various agricultural tasks and domestic duties, obligations which often keep women busy from before sunrise to long after sundown. She describes different patterns for Segou women of varying ethnic backgrounds and she also contrasts women in traditional villages to those in government project areas.

The strongest concern in this paper is about the progressive undermining of the balance of power and independence between men and women. In this concern, she echoes the first paper by Dr. Cloud. Mme. Thiam writes:

> Until independence, women controlled interior commerce almost entirely by themselves. . . . Thanks to this commerce, they had undeniable financial power. . . .
>
> But, after independence, the country having developed cash crop farming, the commercialization of these products became the prerogative of the men . . .
>
> (Now) when women have succeeded in developing a business and . . . (it) reaches a certain size, the management . . . passes into the hands of the men. . . .
>
> Thus . . . we pass . . . to a monetary economy in which women are progressively losing possibilities of revenue . . . since the markets are . . . monopolized by men. At the same time, their dependency vis-à-vis men is growing (p. 76).

Dr. Bernhard Venema, like this editor in the second paper, takes the position that modernization (and aid programs) is not uniformly disadvantageous to women. While he and Creevey agree that women often are ignored in development plans and that their work is vitally important to the development of agriculture—especially to the production of food—he stresses the complexity of the family farming system and the changes occurring within it.

Dr. Venema shows that among Wolof women in Senegal, for example, certain modernization activities increased their opportunities for earning cash stipends. He writes: "The general view is that agricultural mechanization results in loss of work for the woman and consequently implies a loss of . . . influence. As is argued by Boserup (1970), Goody and Buckley (1973), . . . women are pushed back to the domestic domain and lose their position as producers. But . . . Wolof women . . . are not pushed back to domestic work. Neither are they left with traditional tools" (p. 87).

But Dr. Venema also shows that agricultural mechnization was disadvantageous to women in some instances. Thus the introduction of a labor-saving grain mill among Wolof women actually reduced their chance for earning a money income as their paid labor in millet pounding work parties was replaced by the machine.

The substance of Dr. Venema's argument, which he illustrates with examples from Senegalese women but extrapolates to Malian women as well, is that there is no one consistent trend but a very complicated process in which men and women separately and together may gain from one change and lose from another. He acknowledges some of the problems resulting from the male-dominated process of agricultural planning but warns against too easily assuming that all that has been done has a negative impact on women.

The disagreement among the four papers is less important than their concurrence on several major points. In the first place, all four authors agree that the agricultural system of Mali and the Sahel is one in which men dominate and women receive inferior access to the resources needed for agricultural production. Secondly, they all stress the importance of women in Sahelian agriculture and the crying need for programs to be devised which will help women work more efficiently and effectively, reducing their labor time while increasing their productive output. In addition, all would argue that programs must be devised carefully by taking traditional methods of production and traditional relationships among men and women into account before proposing any changes.

1 Sex Roles in Food Production and Distribution Systems in the Sahel

Kathleen Cloud

Editor's Note

Dr. Kathleen Cloud is a research associate in the Office of International Agriculture and in the Department of Agricultural Economics at the University of Illinois. She was founder of the Women and Food Information network, a private, nonprofit organization for education and research. Dr. Cloud has visited the Sahel on several occasions and has written extensively on the subject of women and agriculture in this region. The original version of this chapter was prepared for the International Conference on Women and Food held at the University of Arizona in 1978. Dr. Cloud was one of the organizers of this conference.

Overview of the Sahel

The disastrous drought of the 1970s focused world attention on the Sahel and prompted massive international relief efforts. As the drought abated, consensus grew that to prevent such massive human suffering from recurring, a large-scale, long-term international development effort for the

Sahel was necessary. Such an effort is now under way, with participation by UN agencies, the World Bank, the European Development Fund, many individual nations including the United States, and the African nations themselves.

Massive international development efforts will continue to be focused in this area over the next decades. Knowledge of current food production and distribution systems is essential so that improvements to these systems can be made in rational and integrated ways. Various studies of Sahelian food systems have been done, but they have tended to overlook sex role differences in responsibility for food production, food processing, and food distribution. This case study will identify the roles and responsibilities of women within Sahelian food production and distribution systems. When their role is more clearly understood, it should be possible to plan more effectively.

Any discussion of food production in the Sahel must start with a description of the natural environment. The Sahel is a band of land about 200 miles wide, extending across Africa from the Atlantic 2,600 miles inland, and including much of Senegal, Mali, Niger, Upper Volta, Mauritania, and Chad (see Appendix). It is bounded on the north by the Sahara, on the south by a tropical area of endemic disease. There is one rainy season a year in the summer months. The amount of rainfall decreases as one moves north. Two eco-climatic zones are described in the AID Development Assistance Program (Development Assistance Program 1975) for the regions:

> The *Sudan* zone, with 20–40 inches of rain, can support relatively intensive systems of agriculture. Health conditions are favorable here in comparison with the Guinea zone to the south. Over most of the Sudan zone, millet, sorghum, and cowpeas are the principal food crops, and cotton and groundnuts the cash crops. The possibilities for further diversification into crops such as maize and soybeans are substantial and, as pasture growth is better than in the Sahel, mixed farming is possible and in some areas is being developed. A feature of the cultivated areas of the Sudan zone is the type of parkland where scattered mature trees of economic value, e.g., the shea butter tree, which produces a cocoa substitute, stand in cultivated fields.
>
> The *Sahel* (an Arabic word meaning "border" or "shore") receives 10–20 inches of rain annually. A vast area encompassing some 2 million square miles (two-thirds of the area of the United States) extending 2,600 square miles between latitudes 10–20 degrees north, the Sahel is typically an acacia-dominated, tree and shrub savannah.

Crop production is possible in the Sahel: millet is grown under as little as five inches of rainfall, and groundnuts under as little as 16 inches. Not surprisingly, under such conditions, yields can be good but they are unpredictable. Pastoral operations are the zone's most important economic activity, and under more or less normal conditions nomadic pastoralists in the zone maintain an estimated 19 million cattle, 29 million sheep and goats, and 3.3 million camels, horses, and donkeys. For the nomadic grazers, the Sahel represents a base which provides adequate forage for their herds during four to five months of the year; thereafter, the herds move southwards to graze in areas which, while better watered, present disease hazards in the wet season. A substantial number of breeding females and young stock, however, remain in the Sahel in a normal dry season.

To quote further from the DAP:

> This region is one of the poorest on earth. Some 90 percent of the population lives in rural areas, where subsistence agriculture predominates. Few roads are paved, many areas are difficult to reach and some are inaccessible. In addition, the meager capital wealth is concentrated in the hands of a few. Illiteracy rates average 85–90 percent. While the United States is dissatisfied with an infant mortality rate of less than 20 per thousand, countries of this region have rates which vary between 100–200 per thousand. In some countries only one-half of the children born alive can be expected to live beyond the age of five years. Nonetheless, the current growth rate of population is estimated to be 2.2–2.5 percent per annum.

The social systems of the Sahel have adapted to seasonal, annual, as well as cyclical variations in rainfall in a variety of ways that permit considerable expansion and contraction of food production systems. Nomads travel north to graze on open range when the rainy season produces grasses. They return to the wetter south when the harvest is over to graze their cattle on the farm stubble. Farmers plant more and weed more when grain reserves are low. Young men go to work in the coastal cities when times are hard, taking whatever work they can get. Pastoral families usually have a family branch in the richer, moister south who can manage family trade and absorb some family members in the bad times.

In the really desperate times of drought, whole herds of cattle were driven far south into the tropical game reserves of Nigeria. They were kept there illegally and at risk of sleeping sickness in a gamble to save some of the herd. This measure must have evolved as a strategy long before there were governments and borders in the area.

Sahelian societies tend to be conservative and vest authority in older members. Oral cultures have to depend on human memory for successful strategies in problem solving. When times are good, the young may assume it will always be that way. The older people remember the bad times—how to prepare for them and how to survive them. The span between major droughts in the regions may be 40–60 years. In the droughts, the margin for mistake is very small, especially for the nomads. The advice of the old, who have survived previous droughts, is crucial. Food production systems change slowly in the Sahel for good reason. There is a very delicate balance between people and their environment which rests on the experimental wisdom of centuries.

Food Consumption Patterns

Many people in the region are hungry at least part of the time. The degree of hunger depends to a large extent on the presence or absence of rainfall. There are seasonal variations in hunger; food is shortest just as the rainfall begins, when the previous year's crops are most depleted and animals are producing little milk. In a nutritional survey in Senegal, people weighed least just before the first rainfall. Some years are worse than others. If the rains do not come at the right time, or miss certain areas, for instance, many people go hungry that year.

Table 1 and Figure 1 show this seasonal variation in both food consumption and calorie intake over the course of the year in two different areas of the Sahel. Both samples were done before the drought, in relatively good years.

Firm data on the relative amounts of food consumed by men, women, boys and girls are very scarce. An Economic Commission for Africa (ECA) document (United Nations 1974b) gives a descriptive account of food consumption patterns.

> Unfortunately, in many areas, men of the household get the lion's share of available food and in particular the soups, stews and relishes (which women produce—ed.). In some African cultures, it is

Table 1

Seasonal Variability of Food Consumption
South Chad, 1965

	Grams per capita per day			
	3/15–6/15 Hot—Dry	6/15–9/15 Heavy agricultural labor	9/15–12/15 Harvest	12/15–3/15 Cool
Cereals	441	371	332	472
Tubers	36	64	136	105
Oils	48	64	172	61
Starches	70	75	112	50
Legumes	18	103	175	31
Calorie Equivalent	2,295	2,196	2,841	2,493

Source: SEDES quoted in Intech, Inc., *Nutrition Strategy in the Sahel, Final Report.*

still considered ill-mannered for a woman to eat much of the more nutritious foods, in spite of her higher physiological needs. Within households, women are likely to consume a lower proportion of their requirements than men, not to mention children, girls as opposed to boys. (ECA/FAO Women's Unit 1974)

Sahelian Food Production Systems

Most food production and distribution is still within the framework of a traditional subsistence economy. People raise much of what they eat; social obligations and barter provide much of the rest. "In a subsistence economy the result of work is not intended for exchange, but for consumption by the worker or his immediate companions, and the work, of course, is not remunerated. In a money economy, the results of labor are intended for exchange. The work and its wage allow the worker to

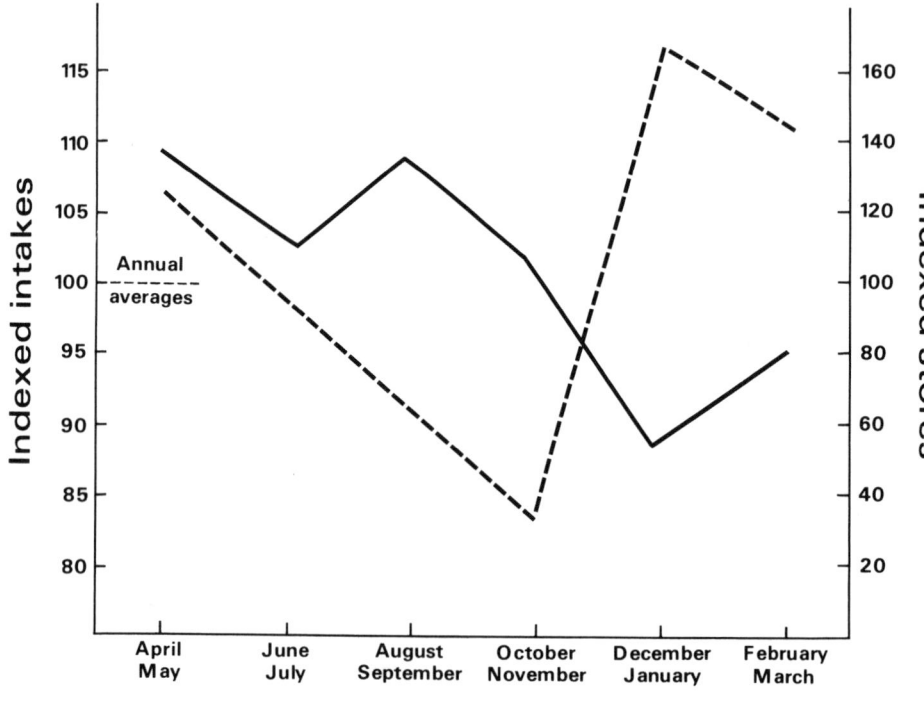

Figure 1.

Seasonal Variance of Calorie Intake in Relation to Stocks of Staple Grain (Guineacorn or Sorghum).

Seasons are defined as bimonthly periods which roughly correspond to the farming calendar.

April/May: Period of planting of millet, land preparation, and beginning of rains.

June/July: "Labor bottleneck" period of ridging, weeding, more planting. Women gather wild fruits.

August/September: Beginning of millet harvest in early August, continuation of weeding tasks and rains.

October/November: Beginning of harvests of groundnuts, rice, peppers, and other vegetable crops, end of rains.

December/January: Guineacorn, cotton, sweet potatoes, and sugarcane harvests.

February/March: Essentially non-farming months.

Village data have not been adjusted for the small sample sizes; all villages are combined without weighting. The index number 100 is taken as the level of average annual intakes and storage.

Solid line: Calorie intake   Dotted line: Stocks of grain

Source: Simons, Emily, "Calorie and Protein Intakes in Three Villages of Zaria Province," May 1970–July 1971. *Samari Miscellaneous Papers* (Nigeria) 55 (1976) p. 25, Fig. 1.

participate in the mainstream of economic activity. Someone who has nothing to exchange is excluded from the mainstream." (Hosbaum 1964).

In the Sahel, small amounts of surplus farm production move into the monetarized sector of the economy, either through the open markets or through government purchasing agencies. Men's work and women's work have different levels of access to the money economy; this fact, which has enormous practical ramifications for development planning, will be discussed more fully later in the paper.

First, a description of the Sahelian food production systems is necessary. For simplicity's sake, the discussion will center on two major types of food production systems: sedentary farmers and pastoralists. These two groups exist in overlapping territories and are interdependent. There are literally hundreds of variations in each pattern; no group displays all of the characteristics of its type, but a general description does serve to give an overall picture. In planning specific projects, it is, of course, important to investigate the specific sex role responsibilities of the groups involved in that project.

Sedentary farmers live in small extended family villages in the wetter areas of the Sahel. Many families are polygamous. They practice slash and burn agriculture that makes good use of their scarcest commodity: labor. Lands are held in common with some combination of inherited usufruct rights, available labor and need determining land assignment. There are five main areas of food production among sedentary farmers: (1) grain production, (2) vegetable gardening, (3) gathering of wild plants, (4) hunting, and (5) small animal production.

Sex Roles in Food Production Among Sedentary Farmers

Grain Production

The grain is usually millet or sorghum. These are most often seen as men's crops, and the husband or group of brothers will control the field and its product. The division of labor is often as follows: Clearing the land is done by boys and young men during the dry season. Trees and large plants are cut down and the area is burned to prepare for planting. Trees with some use (fruit, shade, fodder) are left. In planting, men make holes, women plant seed—often women are responsible for selection of seed from previous harvests to be used. Because of erratic rainfall, they will

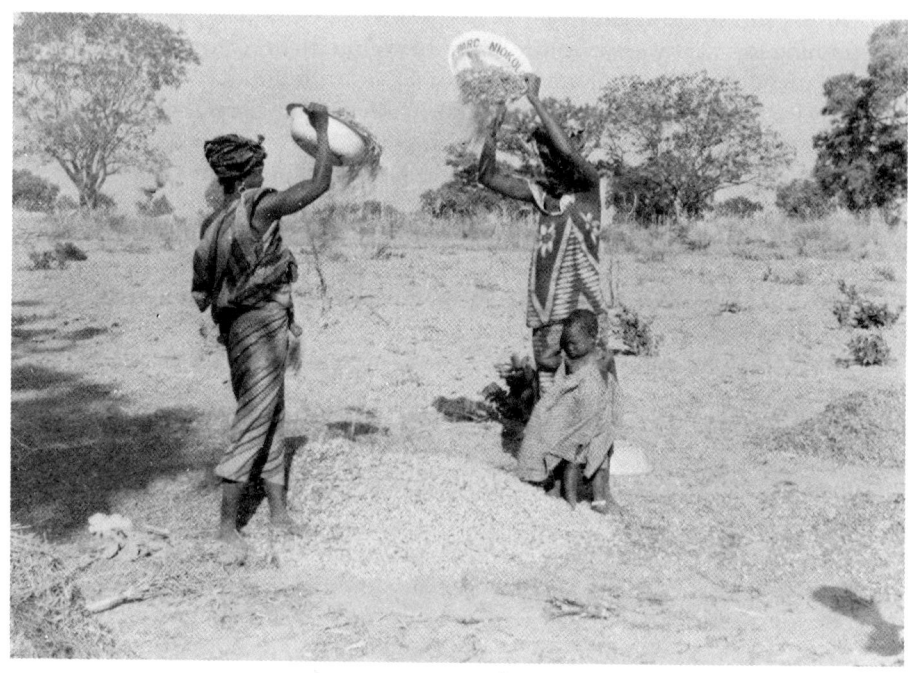

In Senegal, as in other countries in the Sahel, many women participate in all phases of the planting, weeding, and harvesting of crops. In this picture Wolof women are winnowing peanuts in the field. Photo by Bernhard Venema

sometimes plant four or five types of seeds with varying moisture requirements in the same plot. Weeding is the most labor-demanding part of the grain farming, and in most instances every available hand will be used in hoeing weeds. Young men come home from the city to help during this period. Wives will take turns staying home to cook and care for the children while the others go to the fields for the day. A man with several wives and many children has a distinct advantage in agriculture because of the labor he can call upon during the weeding and the harvest. The crops may be weeded one, two, or three times. The amount of weeding has an effect on the amount of grain harvested. There is some indication that when grain reserves are high, less weeding is done—there is not the urgent need for grain. In harvesting, again, every available person will tend to be used. Generally, men are responsible for building the family storage sheds and supervising the grain stored in them. Women are responsible for the household storage of the grain. Threshing is the women's job, and it will be done just before pounding the grain into flour each day. This threshing and milling may take a woman two or three hours, and is one of the most arduous, time consuming tasks she has to perform. There are some exceptions to the pattern of male dominance in grain production. In addition to assisting in their husbands' millet fields, women from some groups will have their own grain fields where they and their children do all the work. Notable among these are some of the Hausa women. In Mali, women grow corn in fairly large quantities and in some areas swamp rice is grown by women.

Vegetable Gardens

Women in most sedentary farm groups have hut gardens where they grow vegetables for the sauces eaten with the millet as well as for trade. They may grow carrots, red peppers, onions, garlic, tomatoes, eggplant, gumbo (okra), and various kinds of beans. It is these sauces that provide the necessary additional amino acids to the millet to make a complete protein chain. In addition, they provide many necessary vitamins, minerals, and fats to the diet while also providing variety in flavor and appearance.

Near urban areas, where there is a cash market for vegetables, they may be grown by men, often with the help of the whole family.

Gathering of Wild Plants and Fruits

This is done almost exclusively by women. In many groups, the gathering of wild foods provides a significant addition to food supplies. This is especially true at the beginning of the rainy season when wild leaves, grass seeds, and fruit provide a supplement to low food stocks. Wild grass seeds are pounded together with millet to add flavor to porridge. Wild leaves are added to the sauces and some of them find their way into the markets, entering the cash economy. Baobob leaves in particular have strong market value, providing cash income for women. Some of these leaves have a surprisingly high protein content as well as furnishing vitamins and minerals.

The importance of gathering wild foods increases manyfold during years of crop failure. This is very important. Wild foods in time of stress provide a most vital reserve. Again, you have the flexibility of systems; if all goes well, people prefer a subsistence farming approach; but if the rains don't come the way they should, the system falls back into its original pattern: hunting/gathering. This, of course, is possible only if some of the traditional bush is available. Projects which eliminate "useless bush" on a grand scale can have terrible effects in that they eliminate the fall-back reserve of the people. All too often visitors see the bush as useless, but in reality there is scarcely a plant that is not used for feeding people or keeping them well (Weber 1978).

Fruit is eaten enthusiastically when it is available. One thing that makes it especially popular is that much of it ripens before the new crops at a time when food supplies are low. A second is that it requires no preparation; it can simply be picked and eaten. However, quantities of fruit are often wasted that, with simple drying frames, could be preserved for the dry season. Several consultants have suggested the introduction of orchards into parts of the Sahel, and in fact, among the Mossi, people do plant fruit trees as a kind of old age insurance "giving people an expectation of minimal income with little expenditure of effort" (Lahuec 1970).

One wild crop of considerable economic importance is the *karite* nut. It is harvested in the summer and buried in pits—later in the fall it is roasted and pounded by groups of women to extract its oil. The karite oil, or shea butter, is then mixed with dough, rolled in leaves, and packed in jars. These balls of oil and dough are either sold in local markets for use in sauces or purchased by wholesalers (men) who refine the oil and export it. In some West African countries it is one of the largest agricultural exports.

Hunting and Fishing

Hunting was at one time a more important food source than it is now; it was one of men's major food producing activities. Big game is gone and smaller game is much scarcer since large areas have become deforested. Game birds, snakes, and animals still provide some protein in Sahelian diets. In many areas, there is a taboo against women and children eating birds, eggs, or snakes, so the protein goes to the hunters. Termites and locusts swarm during the rainy season, and children have feasts on them, roasting them over an open fire. Fishing occurs in rivers, streams and marshes. In some areas it is a major source of protein.

Small Animal Production

Women are primarily responsible for small animals—goats, chickens, sometimes sheep and pigs. They are not raised primarily for meat, but "to make more." Chickens and young animals are kept in the compound. The older animals may be herded by children or kept in corrals while crops are growing. In some places, compost from goat droppings in the corral is used for fertilizer. With proper breeding procedures, goats are a reliable source of milk year around, providing cheese and milk sauce for millet porridge. They recover faster from drought and reproduce more quickly than larger animals. As a result, they have great value as a food source in difficult times. When their milk producing years are over, their meat finds its way into the sauce pot, often at feast times.

Food Distribution Practices Among Sedentary Farmers

In most farming groups, husbands and wives have reciprocal obligations to provide one another and their children with certain things. There is rarely a single household budget in the western sense. Often the husband is to provide grain as well as most meat and fish for the family. The wife is to provide the vegetables or make the sauce that accompanies the grain. She is responsible for preparing both the grain and the sauce for eating, as well as for brewing the beer used on social occasions.

In addition to her responsibilities for the provision of food, the wife, is responsible for the health of the family, often paying for necessary medicines. Both husband and wife may be responsible for part of the clothing needs of the family. The man is responsible for the defense of the family (although since colonial times this obligation has not been as important) and for house-building. The wife, the husband, or both, may be responsible for children's school fees. If there are several wives, each uterine family of mother and children forms a somewhat separate economic unit (Pala 1976).

Under Moslem law, the role obligations are somewhat different. The husband has an obligation to support his wives completely. This is an obligation which only the richer Sahelian Moslem families can accommodate.

In most households, more traditional African patterns prevail. In some polygamous households, the wife cooks each night for her husband; in others, the wives only cook when the husband is to spend the night with her. Denise Paulme, in her introduction to *Women in Tropical Africa* (1963), has this comment on the uses of this obligation:

> The task of preparing the meals is not without its compensations. It provides women with a means of exerting pressure when necessary, as when a man is having an affair to which his wife wishes to raise objections. If he remains deaf to her first remarks, she resorts to a simple method for curing his faithlessness: one evening the husband will find no dinner waiting for him when he comes home. Aware of his guilt, he does not dare to protest and goes to bed with an empty stomach. The next morning he gets up, the same scene is repeated, without a word being said. The husband can do little about it, for if he starts shouting, his wife's complaints will raise all the women of the village against him.

As a general rule, men control the decision making about the disposal of grain crops. Once their family and group obligations are met, they may store the grain or sell it as a cash crop. Women control decision making about surplus vegetables and legumes grown in their hut gardens, and wild plants such as baobob and karite. They will often sell the excess at the market. As a woman gets older and has more children to help her with gardening and gathering, she may have considerable excess for sale, and travel to fairly distant markets, becoming an "own account" trader of some substance. With technical assistance for such women, more fruits and vegetables could enter the money economy, bringing many women a small income. Already, men are moving into the production of vegetables as a cash crop in several areas where an urban market exists. Care needs to be taken that women are not squeezed out of vegetable production for the money economy.

Chickens may be given as gifts or used in ceremonial meals. In some cultures, women sell them in the markets for cash. Chickens and eggs may be eaten by the whole family or just by the men, depending on local taboos.

In fishing villages, women sometimes sell the fish they smoke, and in the coastal areas, some women are fish wholesalers, doing substantial business. Unfortunately, they are gradually being squeezed out of the market as refrigerated warehouse and freezer plants are introduced.

Most of the crops grown specifically for cash (peanuts, cotton, gum arabic) are seen as men's crops, although women often contribute labor to them. They are grown in men's fields and the cash profit goes to men. It is used to pay taxes, to reinvest in farm inputs (fertilizer, better seed), or to purchase symbols of modernity such as radios. Seldom does the income find its way back into the family food budget.

There are, however, some interesting examples of women's cash cropping. In Upper Volta, a UNESCO project[1] has introduced the growing of soybeans as a cash crop for communal women's groups. The proceeds of the sale are used for such things as buying medicine for the village dispensary. Several United States Agency for International Development (USAID) projects in the area also were attempts to help women and women's groups develop cash crops in much the same manner.[2]

Sex Roles in Food Production Among Pastoralists

The second major food production system in the Sahel is that of the pastoralists. They live in small, extended family groups, many of them polygamous. These nomadic and semi-nomadic people have developed movement strategies that permit them to make use of very dry areas for food production.

During the course of the year, animals and people may move considerable distances to take advantage of various food and water sources. Herds and people are combined and recombined in various ways to produce the best conditions for food production with the least stress on animals, people, and environment. The major food production activities are: (1) stock breeding and milk production, (2) gathering of wild plants, (3) hunting, and (4) vegetable gardening and grain farming. Again, sex roles for each activity will be described.

Stock Breeding and Milk Production

Most pastoralists breed a wide variety of animals and maintain diversified herds as an adaption to the environment. Camels, cattle, sheep, and goats each have characteristics that provide different benefits. Goats breed quickly and recover quickly from drought. They can exist on browse when grasses are not available. Both their milk and their meat are palatable. Sheep give somewhat more milk and their meat is considered tastier, but they are more vulnerable to drought than goats, and herds take longer to reconstitute. Both sheep and goats stay fairly close to camp. They are herded by boys and girls and are milked by women.

Cattle can go further from water for pasturage than either sheep or goats, and when they are fresh, they give considerably more milk. Cattle are taken on long treks to the north during the rainy season by boys and young men and return after harvest to graze on farmers' stubble or fallow fields. Cows with young calves are often left near the camp and milked by the women. Whether men or women milk cattle varies from group to group, but even when women don't do the milking, the milk is seen as belonging to them.

Camels have the largest grazing reach because they can go furthest from water. They produce slowly, but give high quality milk for long periods. Males are castrated and used for transportation and trading; females are used for breeding and milk. Camels are the exclusive responsibility of men, even the milking. Not all pastoralists have camels; some prefer horses or donkeys for transportation.

Gathering of Wild Plants

Among pastoralists, collecting is mainly the task of women, but boys may also participate. Among the Tuareg, more than 50 different plants are gathered: seed, leaves, or fruit. As an example of the volume of this production, one Tuareg household gathered 1,000 kilograms of *iceben* (small wild grass) seeds in one season. They are pounded along with millet to give flavor to the porridge. The leaves are used in sauces, just as in farm families. Fresh fruit is consumed with enjoyment.

Hunting

Hunting of small game is still sometimes done by men, but here also it is not as important as it used to be because there is less game.

Vegetable Gardening and Grain Production

Some few nomadic groups farm around oases in the northern Sahel. They use irrigation, raising water from shallow wells with a bucket and a pole or animal traction. They grow wheat and some barley in the winter, millet and sorghum in the summer. Tomatoes grow most of the year. Potatoes, sweet potatoes, onions, melons, dates, and sometimes lemons, beans, saffron, red peppers, and mint are each grown in the same areas. Millet and sorghum are harvested by women, dates by men, and other crops by both men and women.

Food Distribution Practices Among Pastoralists

Among pastoralists also, husbands and wives have reciprocal obligations to provide one another and their children with certain goods and services. Again, there is not a common household budget in the Western sense. In general, women are responsible for the provision of household goods, pots, chests, utensils, and for the processing and trading of milk and milk products. Men are responsible for the care and herding, as well as the actual selling, of the large animals, although they may not be their owners.

The ownership and usufruct rights to nomadic animals is one of the murkiest areas of knowledge in development planning in the Sahel. The general assumption of development planners repeated to us all over the Sahel was that men owned the cattle; women *might* own goats and sheep. However, an examination of the literature, including the AID-sponsored Rupp (1976) report, shows this to be a misconception. Animals are owned by individuals, but herded as a group responsibility. According to Nicolaisen (1963), among the Tuareg almost everyone is a stock owner. Even little boys and girls may own a few animals which are given them by their parents or close relations. Offspring of these domestic animals also belong to the children, but the milk, butter, and meat should serve the needs of the household to which they belong. Within the household, the husband and wife also have individual animals. Among the Tuareg, the husband or wife can freely sell or slaughter animals they own without asking permission of the spouse, while among the Fulani they must consult before selling. In both cases the meat or money should serve the needs of the household.

A woman may have title to animals in two diffeent ways, with different arrangements for their management and disposition. First, there are the *bridewealth* animals, paid by her husband's family, to her father or oldest brother, but the offspring of these bridewealth animals go to the bride or her children. These animals are kept with her father's herds and her maternal family has use of the milk or the meat if they are slaughtered, but the offspring continue to belong to her uterine family. Among the Tuareg, if there is a divorce the bridewealth is not returned but is used to provide for the children.

A second kind of ownership is more directly under the women's control. It is the obligation of the bride's family to send her to her new home with a dowry consisting of household goods and animals—usually five or six donkeys and between 10 and 40 goats. Sometime after marriage it is customary for a husband to give his wife a gift of animals according to his means—a few goats, one or two camels. This gift remains in her husband's camp so that the animals serve the needs of his household and their offspring.

In Madame Rupp's seminars with both Fulani and Tuareg herders, one of the major concerns expressed was that the government's program to reconstitute herds lost in the drought was replacing cattle only for the men. Women's stock was not being replaced. This was crippling their social system—animals were unavailable for dowry and bridewealth payments, and women had lost their independent property. This was apparently the unintentional result of the government program that issued a card to the head of each family and gave animals only to the family head. Program administrators' lack of understanding of sex roles in the control of resources seriously damaged nomadic women's economic and social positions.

Because usufruct rights are important among the nomads, people who are in need will be given the use of animals temporarily. Families also have rights to the use of animals they do not own, such as the cattle of sedentary farmers taken north in the collective herds during the rainy season.

The disposition of the milk and cheese that is a product of all these animals is the woman's responsibility. When and where it is possible, she will trade milk for millet from sedentary farmers. In good times, the trade ratio is a measure of millet to a measure of milk. If times are bad for one group or another, the ratio will change. Sometimes the pastoralists will exist entirely on milk for months. One source (Galon cited in Nicolaisen 1963) cites four liters per day as the necessary amount. Nicolaisen himself cited 8–10 liters per day. Nomads say they get "weary" from just milk and prefer other foods.

Men trade further afield and use the cash profits to buy grain. In many groups, the men have traditionally been traders and middlemen transporting goods for long distances. These trading caravans have diminished in importance, and are no longer a major source of income for most groups, but men still trade animals vigorously. In some cases where nomads have settled near towns and cities, milk has cash value. When milk is sold for cash, the trading sometimes passes out of women's hands and into men's.

I could find no indication that vegetables, cereals or gathered food were produced in large enough amounts by pastoralists for surplus to be sold. Their major cash product, usually sold by men, is meat and occasionally milk.

Female goats, sheep, and cattle are all slaughtered for food sometime before the end of their reproductive years, often for ceremonial occasions. The meat is consumed by the family or the live animals are sold for slaughter. Younger bulls and bullocks are sold to traders and are the major cash crop of the pastoralists.

Sex Roles in Food Processing

Between the time food is produced and consumed, most of it has to be processed in some way. Sometimes this is done before distribution, sometimes after distribution. Since there is such commonality in the patterns, for simplicity's sake, I will discuss all food processing activities for both farmers and pastoralists together here.

The major food processing activities shared by both groups are water carrying, both for drinking and sanitation, cooking, including the gathering of wood and making of the fire, threshing and pounding of grain before cooking, and drying and processing of foods for storage, such as fruits and vegetables, baobob leaves, and karite oil. In addition, farm women are responsible for brewing beer for social occasions and nomadic women are responsible for processing milk into cheese and butter. All these food processing activities are done exclusively by women, and almost all of them are subsistence activities. With the exception of some processing of milk and karite oil, and the drying of wild leaves, none of these activities produces any money.

These activities consume major portions of women's time and energy. Food could not be consumed if these activities were not performed, yet they often are invisible in accounts of food systems. Economists do not generally include such activities in their accounting (Spencer 1976) and as a result, development planning tends to overlook these activities. Thought needs to be given to ways of making these activities more visible within the planning process.

One solution to this problem is to look at the labor involved in

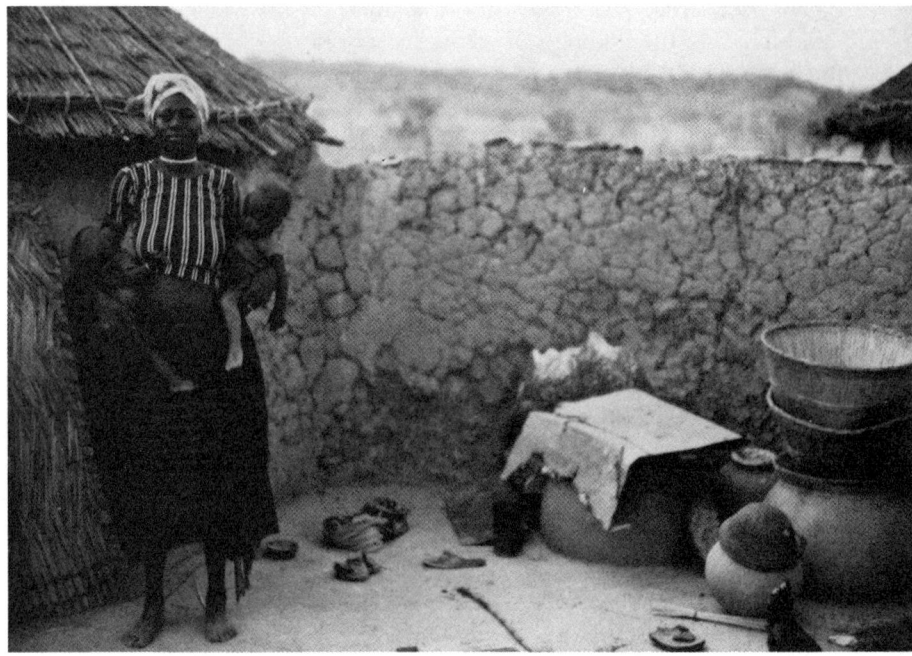

This pregnant Mossi woman with her two small children stands outside her hut in the village of Koukoundi in Burkina Faso. To her right is equipment for making traditional beer. Photo by Helen Henderson

various food-related activities, and to use a measure of labor as a way of making women's contribution more visible.

The ECA report on women's participation in food production and processing activities (United Nations 1974b) uses the *unit of participation* . . . for measuring women's labor in rural Africa. "To obtain a unit of participation . . . one makes the best estimate, based on available data and experience, of the percentage of labor associated with a particular task which may be attributed to women and express it as a fraction of 1. For example, it is estimated that in Dukohata, Tanzanian men work 1800 hours per year in agriculture and women work 2,600. This totals 4,400 hours of which 60 percent is women's work. Women's unit of participation is thus 0.60." Using the method, they attempted to arrive at rough estimates of the participation of women in the traditional rural and early modernizing economy in Africa as a whole in order to provide a model.

African Women's Participation in Food-Related Activities

Production/Supply/Distribution	Unit of Participation
1. Food production	0.70
2. Domestic food storage	0.50
3. Food processing	1.00
4. Animal husbandry	0.50
5. Marketing	0.60
6. Brewing	0.90
7. Water Supply	0.90
8. Fuel Supply	0.80

The ECA report suggests that research needs to be done which would permit units of participation to be determined accurately for areas within countries, then on the national level, then for all of Africa. A limited amount of such research is included in several projects going on currently in the Sahel. Much more of this research needs to be done to provide data in quantitative as well as descriptive terms. My own impressions of the division of labor in the Sahel lead me to suspect that the figures for food production and marketing might be slightly lower than the African averages cited in the table, but only sufficient research can establish what the proportions actually are.

Summary of Women's Roles in Sahelian Food Systems

To summarize this description of women's traditional roles in food production, preparation, and distribution in the Sahel:

1. A significant amount of food production is accomplished by women, primarily in the areas of vegetable growing, gathering of wild plants, small animal production, milking, and the processing of milk products.

2. Almost all food processing is done by women. This includes threshing and milling of grain, cooking, drying and preserving of fruits, vegetables, and leaves, brewing of beer, and making of cheeses and butter, as well as the gathering of firewood and transportation of water that are necessary for these processes.

3. Most of the food produced and processed by Sahelian women is consumed by "their immediate companions" within the subsistence sector only. A small portion of women's food production reaches the monetized sector, usually the local markets.

Recent Changes in the Sahel

The foregoing had provided a description of women's traditional roles in Sahelian food systems. These total food systems were affected first by some degree of modernization and then by the drought. Currently, there is an attempt to affect these systems in a planned, rational way through long-term development programs. I would like to describe briefly the impact of each of these on the systems.

Modernization has not penetrated very deeply into much of the Sahel. The French pacified the nomadic tribes that raided the area, and modified the feudal relationship they had had with sedentary farmers. Some endemic diseases were brought under control for both humans and livestock, thereby increasing population growth rates. Cash crops for export were introduced and men began to farm them in small plots, but there were only a few of the plantations similar to those which developed in other parts of Africa. Plow agriculture was expanded. In the early 1960s, there were a number of deep wells bored in the north to carry the cattle through dry periods. The French educational system was introduced, and while a small number of Africans went straight through the system and into the best French universities, most of the population was left untouched. To quote from a report describing the years just before the drought:

> Human population pressure continued to rise and export crops became an important part of the output, replacing traditional culture in more favorable areas. The resulting pressure for increased production decreased fallow time and lowered productivity per hectare, even though total production continued to rise as a result of a larger percentage of the land being used for agricultural activities in any given year. Further, the expansion of cultivated lands in the moist areas decreased available grazing lands. Thus, even greater pressure was placed on the exceptional forage productivity of the Sahel. Heavy cutting of trees for firewood near urban areas contributed to ecosystem destruction. (Matlock and Cockrum 1976)

For a while, the system continued to be able to handle the pressure because of very high rainfall levels in the 1960s. But then the rains diminished, and in 1972 and 1973 many areas had no rain at all. The drought's impact was quick and dramatic. According to the area Development Assistance Program:

The U.S. Center for Disease Control in Atlanta undertook a nutritional survey in 1973 which estimated that as many as 100,000 people may have died. International experts have estimated that perhaps 40 percent of the goats, sheep, cattle and camels on which much of the economy and social structure rests, have fallen victim to the drought, either through death, premature slaughter or early sales. The drought has had a profound effect on the region; a fundamental weakness of the ecological base, disruption of the social and economic relationships and the changing of basic ways of life. (Agency for International Development 1975)

The drought called forth large-scale relief efforts, followed by the institution of international planning mechanisms for long-term development of the region. The international planning group, known as the Club du Sahel, has developed what is in many ways a model of sensitive, rational, development planning for the area. The theme is intensive rural development. According to an AID planning document:

The region is poor in energy and mineral resources. There will be little opportunity for industrialization until agricultural development is assured. The proposed program must not result in energy dependence. Most of the people are rural and their socio-economic basis is in agriculture. The Sahel Development Program will not disturb this basis; the future of the Sahel clearly hinges on its agricultural production framework . . . The region's increased income will work to the advantage of all its people. (Agency for International Development 1976)

The Sahel Development Program[3]

The basic elements of the program are listed as Human Resources Projects, Near Term Rural Development Projects, Far Term Water Basin Development Projects, and Health Resources and Transportation Projects. Because they relate directly to the topic of this paper, I would like to examine two elements—Near Term Rural Development Projects and Far Term Water Basin Development Projects—in more detail. The Near Term Development Projects are intended to provide simple inputs to current farming and pastoral systems to make them more productive. These inputs might include fungicides for seeds, improved varieties of seeds, locally produced fertilizers (i.e., manures and phosphates), and improved crop rotation methods. In some places they would include the introduction of draft animals and plows to relieve the labor constraint in food production. Planning is being done with pastoralists for better placement and management of wells, and methods for managing the rotation of grazing land and delivery of simple preventive health services. As much as possible, the programs are attempting to use the people from the village in the planning processes out of a conviction that they know what their constraints are far better than anyone else.

Far Term Water Basin Development Projects are much more ambitious efforts to utilize the potential of the large river basins in the area with their fertile land and abundant water. Before these lands can be settled, their endemic diseases, such as onchocerciasis and sleeping sickness must be eliminated. Large scale efforts to do this are now in progress. If the basins can be resettled and brought into productivity, their use will provide a basic food supply for the region both in wet years and in dry. Their production, added to the production of the traditional systems, would provide enough food for the expanding population.

This is the way the program is conceptualized. How is it being implemented? In what ways is it responding to women's position within this agricultural production framework?

In some ways it is doing fairly well. There are a number of Women in Development Projects within the region that are bringing work-reducing technologies to village women. For example, cooperatively owned gasoline mills for grinding millet are being distributed through UNICEF in Senegal, UNESCO in Upper Volta and AID in Mali. Some women's cash cropping of vegetables is being done under European funding in Senegal, and American funding in Mali. There are non-formal literacy programs directed to women in the UNESCO project, and at Operation Riz-Segou in Mali, among other places. In Senegal, the government is going through an administrative reorganization, and Village Councils of both men and women are being allocated some funds to implement their own development projects. UNICEF and *Animation Feminine* are working with the women in the reorganized villages to develop small projects—some of these are also bilaterally funded. *Animation Feminine* in Niger has had *animatrices* in more than 200 villages working with women in agricultural production as well as health services. Sometimes they have been able to act as liaison between the local women and a large project to encourage the provision of services to women. In one such case, the European Fund for Economic Development 3 M Project, they were able to persuade the

Project to train the women in animal health and the treatment of seeds with fungicide.

But if the first principle of development is the Hippocratic principle, "to do no harm," then there is a problem. At the same time that some programs are being developed to be responsive to women's needs, other programs are undercutting women's traditional roles by ignoring them. Most of the larger programs seem structured on the assumption that all farmers and pastoralists are men, that all decision making is done by men, that all resources are controlled by men and therefore, a development project staffed completely by men, with male extension workers dispensing training credit and resources to men, is an appropriate program structure. Exceptions to this pattern are far too few.

Pattern of Resource Allocation to Men in Africa

This problem of allocating resources primarily to men is not restricted to the Sahelian programs, of course. Its prevalence as a world pattern has been amply documented by Boserup (1970) among others. It is, however, somewhat more dismaying in Africa with its well-documented gender specific division in social and economic roles. Traditional African societies tend to have two spheres of power, male and female. Sometimes the male power is conceptualized as formal power and the women's power is personal power, but often women's power is also formal and acknowledged. In many traditional societies, a queen mother or a queen sister represented women's power at the top of the authority structure in roles that emphasized the importance of both sexes. Market women's associations, women's age grade groups, wives' associations, and lineage groups all are features of many African societies.

Hafkin and Bay (1976) attributing the modern neglect of the dual sex power distribution to colonial rule in which men had all the power, write, "Traditional systems of dispersed and shared political authority had no place in the colonial system."

Another cause of this neglect may be that much of a woman's food production is for her family's use and doesn't reach the monetarized sector of the economy. It doesn't get into national production statistics, but people are eating it. By starting with the consumer, with what people are eating, a different picture of food production emerges than if GDP or aggregate figures of production for the country are used as indicators. This difference in perspective is crucial in analyzing women's contributions to food production, particularly in subsistence economies.

A good example of the problem is provided by the AID projected budget for 1978 in this region. Of the $32 million budgeted, $24 million went for food and nutrition activities, $5 million for health and population activities, and $3 million for education and human resources. Of the $24 million for food production, the overwhelming amount went to cereal and cattle production, which are primarily men's crops in the monetarized sector. A small percentage went to vegetable production. There is one small project for goat production included. There is no money for chickens, pigs, fruits, or other gathered crops such as shea butter or baobab leaves. Nor are there any funds for milk production or processing. There are, however, small projects for men gathering wild honey in Chad and Upper Volta.

No one seriously proposes that the Sahelian diet should consist only of grains and meat. Everyone expects that vegetables, fruits, greens, milk, and cheese will continue to be produced. It is simply that little AID money is being expended to assist in their production.

Access to Land for Women

Another factor contributing to neglect of women's roles in food production is the fact that much of it takes place on uncultivated land—in gathering, small animal production, and milk production. Alternatively, vegetable gardening takes place on very small plots. It is one of the characteristics of gardening that a great deal of food can be produced in a small space, but this very characteristic tends to work against women. For example, "Cereals are the major crop; many varieties are grown on about 65 percent of the cultivated land . . . Peanuts and cotton occupied about 25 percent of the cultivated area. Small amounts of manioc, yams, sugarcane and tobacco were produced on the remaining ten percent of the cultivated land" (Matlock and Cockrum 1976). Women's crops are invisible in this account of land use although production may be far more significant. This invisibility may also contribute to the lack of development resources available for some kinds of food production.

The question of land use and access to land becomes crucial in areas where plow agriculture is being introduced, particularly in river basin resettlement. As farming practices are intensified and more effort and energy is put into each plot of land, land ownership tends to move from communal ownership with usufruct rights over the land to private ownership. This shift in the control over land is often triggered by population pressures. The increased demand for food produces an intensification of land use. The intensified use of land for cash vegetable production near urban areas is an example of such a shift. This is precisely what is intended in the river basin resettlement projects and the process presents a real threat to women unless it is handled very carefully. Women's current food production activities use very little cultivated land, and most of their products do not enter the money economy. As a result, their existence tends to be ignored by planners. In resettlement schemes, land is often subdivided and assigned to families. The head of the family is the person listed as responsible for repayments. As the land passes into private ownership, it is the family head who has ownership rights, and the rest of the family become his dependents. Thus, as land passes into a more privatized kind of ownership, women are squeezed out of independent access to land. The results for women's power and status are so disastrous that a number of writers (Boserup 1970, Sacks 1975) have identified this loss of independent access to the means of production as *the* development event that marks the marginalization of women.

African women have resisted this marginalization, quite vigorously at times. The famous 1929 women's wars in Nigeria are examples of such resistance. But the process goes on. Within the Sahel there is a current example of river basin settlement that illustrates this problem quite clearly. The French have a project to assist in the development of the White and Red Volta Valleys in Upper Volta. By 1974, 187,000 hectares had been mapped, 1,000 had been cleared and plowed. In 1974 there was space for 250 families to settle, and it was expected that 600 more families could be received before the start of the 1975 rainy season (Moton, G. 1974).

The first families moved onto the land as planned and there was a substantial waiting list for upcoming farms. But within a year there were problems—wives were leaving, families were threatening to move out, and new families were reluctant to move in. The Project Management approached the Voltaic Research Institute to find out why. The answer— because of the required land use pattern, women had no place for their vegetable gardens. The wells were far from the houses, making water for domestic use difficult and time consuming to procure. Finally, the women were not able to care adequately for the family's health because they could not find the necessary medicinal herbs and plants on the cleared land.

Some measures are being taken to correct these conditions, but the more serious questions of long-term private ownership of land has not been addressed.[4]

How to avoid marginalizing women economically at this point is not at all clear. Have any societies passed through this stage with women retaining a measure of control over access to land? If so, what were the conditions of such a successful transition? Are there any alternatives to continued access to land that would provide women with independent resources and some independent economic base such as they have in more traditional societies with usufruct rights and dowries? These are all questions that urgently demand investigation before planning for river basin resettlement proceeds much further.

Whatever the causes, the pattern of exclusion of women's productive activities from access to development resources plagues many of the current development projects in the Sahel. Women's work, women's productivity and women's control of resources are often being denied by the refusal of projects to relate to them.[5]

Attitude of African Women

Governments and development projects are male-staffed. They relate most easily to formal male power structures. One solution to this problem might be institutionalization of visible, formal women's organizations for governments and projects to relate to in systematic ways. Interestingly enough, this organizational visibility is what African women themselves say they want. At the 1974 Regional Seminar on Women in Development in Addis Ababa sponsored by the UN Economic Commission for Africa, African women adopted a Plan of Action similar to the one American women recently endorsed in Houston. In the first resolution they call for a series of organizational structures on the national level that would include: (1) national commissions on women and development to make policy recommendations and action proposals, (2) women's bureaus of permanent secretariats of these national commissions to undertake research, to formulate projects and programs, and, in general, to seek women's integration in all sectors of social and economic development, (3) an interdepartmental body of experts . . . to ensure coordination of programmes and adequate representation within national policies and planning, and (4) a non-governmental organization coordination committee, which might assist women to seek representation in decision-making bodies, to work toward changing attitudes, to supplement public resources, and to promote international collaboration and exchange.

On the African regional level they called for an Africa Regional Standing Committee and a Pan African Research and Training Center to assist governments and voluntary agencies in strengthening the roles of women in the Africa Region. Since 1974, these National Women in Development Commissions have been formed in most Sahelian countries. In addition to these commissions, numerous other formal women's organizations exist at the national level in various countries. Some of them have organizational units that stretch down to the village level. In Senegal, the national political party has a very active women's section that is running training and development programs in many regions. In Mali, Niger, Upper Volta, and Mauritania, there are National Women's Federations and some of these have published policy statements on the very specific development needs of women in their countries. Within the governments of Niger and Senegal, *animation feminine* programs organize village women to articulate their needs and help them to meet those needs at the local level. These groups and others like them need support for expansion. They also need greater visibility to donor agencies. During the summer of 1976, I visited many of these women's groups as a member of a CID/Arid Lands/AID team investigating the impact of development projects on women. Again in 1983 I discussed these problems with women's groups in Mali and Senegal. The women were most eager to share their ideas with us. They have a clear perception of their situation and their needs, and very precise notions of what would be an immediate benefit to rural women.

What did these women's groups say they wanted from the development community? Very simple, practical things.

1. *Relief from the enormous burden of work for poor women.* UNESCO did an initial survey of women in their project area in Upper Volta. The most common request of the women was for relief from their excessive work load. First and foremost, they wanted gasoline or diesel-powered mills for grinding their millet. "Diesel-powered mills work and women want them," (Nariama Wani, *Animation Feminine,* Niger); "where there is a mill women use it" (Louisette Alzoma, Secretary, Federation of Nigerian Women). They also want better access to wells to relieve their work. In some places they asked for pumps for raising water or some way of keeping the water clean in the well. "Men will support labor-saving devices and help dig wells as long as they don't threaten the traditional role division" (Jeanne Zongo, President of the Federation of Voltaic Women).

Two-wheeled carts (charettes) for transporting water and firewood were also mentioned repeatedly in Upper Volta. In Niger, where *Animation Feminine* has had ten years of experience with the village women, their requests were more sophisticated. In addition to mills and improved access to water, they want weeding tools, fungicide for treating millet seed . . . and some Hausa women want animal traction for plows!

2. *Help with gardens* was requested. The village women ask for different varieties of seed, more seeds, and better kinds. In Senegal, Catholic Relief Services had some small women's co-ops working with very simple drip irrigation techniques to extend their vegetable production season further into the dry season. Other women had heard of UNICEF's work with homemade cisterns and wanted help with them.

3. *Help with food preservation,* particularly ways of drying fruits and vegetables and smoking fish. (This last request was from Senegal).

4. *Help with small animal husbandry.* They want information on disease diagnosis for animals, and better information on animal nutrition. The Hausa women want to know about diseases in cattle. They also want better breeds of chickens and goats that they can crossbreed with their own. The Nigerian Women's Federation, in their policy paper, specifically request a particular breed of goat, "La Chevre rousse de Maradi." Upon further investigation, I found this explanation, "among goat breeders of Niger, the Red Maradi occupies an exceptional place on account of its skin, considerable numbers being exported. It is an excellent source of milk and meat while its skin is a source of revenue for farmers" (Robinet 1967).

All four cases above repeatedly mentioned the need for access to paraprofessional and professional training for the women staffing the various programs. In Senegal they needed training in food preservation techniques; in Niger, training in animal husbandry and agriculture. One problem that was repeated to us over and over again in many contexts was the lack of adequate training facilities for women in agriculture, animal husbandry, and rural development within the region. "There is a school for male agents (IPPR) that is being enlarged but there are still no places for women. No institutions are training women in agriculture . . . we would welcome it if you can help us with the training of our agents," (Nariama Wani, *Animation Feminine,* Niger).

In the planned expansion of agriculture training facilities in the Sahel, some slots are to be provided for women, but it would be useful to make a systematic assessment of the needs and the opportunities to see how well they match.

On the question of the acceptability of American women coming over to give technical assistance, "Religious leaders are reassured if women come to work with women" (Mme. Marie Anne Sohai—member, Chamber of Deputies, Senegal).

The final area was the one mentioned most often:

5. *The need for cash income.* "Women need a source of income. They can grow tomatoes, salad, make crafts," (Josephine Gisseau, Upper Volta); "Women need cash," (Barbara Skappa, Peace Corps, Mali). "Women need supplemental income—here in the Center they learn sewing to sell," (Halimatou Orseini, PMI Clinic, Niger). "For rural women it is very important to give them some opportunity to earn money. It will help those who earn and those who don't but know they could. Their families will respect them more" (Boserup 1976).

Literacy and health needs, although outside the purview of this chapter, were also mentioned frequently.

Conclusion

From the foregoing discussion, what conclusions can we draw?

1. Women are a major element in the food producing, processing, and distribution system in the Sahel. Studies should be done to substantiate this contribution.

2. There are several successful projects in the Sahel that are specifically focused on supporting the effective participation of women in these systems. Several more such projects are in the planning stages now.

3. There is a major international development effort going on in the Sahel that is in many ways a model of thoughtful development assistance. However, in spite of some good faith efforts, the presence of women as an integral part of the agricultural system is being ignored in most of the larger projects. Assistance, training, and resources are being delivered to men and men's crops proportionately far more than to women and women's crops. This differential input tends to undercut women's traditional roles and power. Analysis of sex roles and responsibilities of the target population should be included in planning each project so that services and resources are delivered to the appropriate people.

4. The emphasis on development of a few food crops at the expense of others is a poor strategy for assuring adequate food for all. In an

economy where most food for most people will be produced and consumed within the subsistence sector for some time yet, it would be wise to attend to increasing the productivity of a large range of subsistence activities.

5. Intensification of land use, with its accompanying changes in access to land, presents a threat to women's traditional roles and status unless it is handled very carefully. I would echo Pala's point that "research in land tenure changes and women's rights is important and could be profitable . . . What is the impact of land privatization or nationalization on women?" (Pala 1976).

6. A contributing factor in the neglect of women's participation in Sahelian food systems is that women are not present in either African or American government agencies dealing with agricultural development. With few to act as advocates for women's fuller participation in projects, it tends to be ignored.

7. This problem is intensified at the international level in the Club du Sahel.

8. Official government commissions on Women in Development and other official women's groups do exist in many Sahelian countries. They are new and often have a good grasp of the fundamental realities of development, and they are most eager to be involved in the planning of development assistance.

These African women's groups could be involved as a resource in planning integrated projects as well as in projects specifically focused on women.

Notes

1. This was the project on "Equal Access to Education for Women and Girls."
2. AID projects included "The Upper Volta Women in Development Project", 1977–1982, and a project on the "Training of Women in the Sahel", 1978–1982.
3. This is a multilateral program with African and donor countries involved. It is still functioning in the Sahel.
4. Women were allowed to have private gardens on part of the land and some wells were placed closer to the houses. This project will be analyzed in a forthcoming case study by the Office of Women in Development, USAID.
5. See analysis of the impact of eleven AID projects on women in Cloud 1985.

Interviews

Alzoma, Louisette. Secretary, Federation of Nigerian Women. Niamey, Niger. July 14, 1976.
Boserup, Esther. Economist. Wellesley, Massachusetts. June 1976.
Bingham, James. Political Scientist. Bamako, Mali. July 13, 1976.
Compase, Scholastique. Director, UNESCO. Project on Equal Access to Education for Women and Girls. Ouagadougou, Upper Volta. July 1976.
Digne, Ana. Director. Promotion Feminine. Dakar, Senegal. July 26, 1976.
Orseini, Halimatou. Assistante Sociale, Direction des Affaires Sociales, Niamey, Niger. July 15, 1976.
Pala, Achola. Anthropologist, Institute for Development Studies, Kemp. Washington, D.C. April 1976.
Skapa, Barbara. Assistant Director, Peace Corps. Bamako, Mali. July 12, 1976.
Sohai, Marie Anne. Member, Chamber of Deputies. Dakar, Senegal. July 25, 1976.
Wani, Nariama. Agent technique d'animation. Niamey, Niger. July 14–15, 1976.
Weber, Fred. Development Consultant. Niamey, Niger, July 15, 1976. Tucson, 1978.
Zongo, Jeanne. President, Federation of Voltaic Women. Ouagadougou, Upper Volta. July 22, 1976.

2 The Role of Women in Malian Agriculture
Lucy E. Creevey

Editor's Note

Dr. Lucy Creevey is professor of City and Regional Planning and director of the Program in Appropriate Technology and Energy Management for Development at the University of Pennsylvania. Since 1965, she has worked frequently in the Sahel on research projects and as part of planning teams focusing on increasing village food production. She was the CILCA organizer and *rapporteur* for the Workshop on the Training and Animation of Women, Bamako, Mali, 7–9 June 1983. This chapter attempts to make some of the general points raised by Dr. Cloud specific to the situation in Mali. Dr. Creevey, however, differs from Dr. Cloud in her emphasis, as she stresses the interdependence and complexity of the roles of men and women in the process of change whereas Dr. Cloud emphasizes the progressive loss of status and power of rural women during modernization.

A rural woman participates in virtually all the work in the fields. It is not even rare for her to undertake this work alone if the men are absent.

Along with the family field, the woman sometimes has her own personal plot, her kitchen garden. She draws the water for the household needs and for watering the animals, she goes to fetch wood for the kitchen, she searches the bush to collect the wild [herbs and spices]. She goes from one market to another to sell her products and to buy condiments and other commodities and articles of prime necessity for the family needs. She has charge of domestic work. She must also take care of the heavy obligations of a wife and mother.

It is impossible to make an exhaustive list of all done by a rural woman who scarcely has time for rest. Her many tasks never let up. She is the center of production and of the fight for survival.

Aly Cisse 1983[1]

Introduction to Mali

Mali, a nation of 7.1 million people located in West Africa (east of Senegal, north of the Ivory Coast and Upper Volta), became independent in 1960 only a few months after it had received independence from France as part of the Mali Federation (see Foltz 1965). Mali's president, Modibo Keita, was a strong leader, head of the political party Union Soudanaise, which had managed to unify many diverse tribes across a vast land area (1,240 square kilometers) (see Morgenthau 1964). But in the period following independence he faced economic chaos. Once having broken with Senegal (its partner in the Mali Federation), Mali was landlocked, its railroad tie to Dakar broken off. The countries with which it had the closest ideological ties, Guinea and Ghana, were situated so that trade was impractical and realistic collaborative development plans virtually impossible. Keita, a socialist, moved even further left in his efforts to find a way out of the situation he faced. In 1962 he withdrew Mali from the franc zone. He tried to nationalize business and banking, to force rural producers into cooperatives, and to attract assistance from the Communist Bloc. He began to strictly limit the activities of his critics whose allegations became increasingly severe as Mali plunged further into economic crisis.

In 1966, parliamentary government was suspended and Modibo Keita assumed an increasingly dictatorial role in Mali. Since this action was accompanied by economic stagnation and increasing hardship for Malian citizens, it is not surprising that, in 1968, a young lieutenant of the armed forces, Moussa Traore, led a coup which removed Keita from office (see Morgenthau and Creevey 1984).

The new military government confronted a nation in which no major mineral resources had been discovered and which had received very little investment from its former French colonial rulers. Even Modibo Keita had attempted to return to the franc zone after his earlier disastrous withdrawal had made the Mali franc impossible to convert into other currencies. Traore now made strong efforts to go further to win back support from the French and from other western allies, notably the United States, to try to bring in investment and loans so that development could proceed.

In the early 1970s, Mali was subject to a killing drought which brought the country to the forefront of international attention. Vast numbers of cattle starved and thousands of people died. International assistance took the form of food and medical aid while the government desperately sought ways of helping the much neglected and extremely impoverished agricultural sector. During these difficult years, Traore maintained a military regime, but in 1974 he took the first steps to return to civilian government by approving a new constitution. In 1976, the Union Démocratique Populaire de Mali (UDPM) was formed, a single party to group all people in the nation in one institution. It was to be a "non-ideological, democratic union." In June 1979 elections were held for the legislature and for the president. Moussa Traore was elected by more than 99 percent of the vote.

The government of Mali is organized into sixteen Ministries including Rural Development, Economy and Planning, Foreign Affairs and International Cooperation, Agriculture, National Education, Public Health and Social Affairs, Information and Telecommunications, Equipment, Sports, Arts and Culture, Transportation and Public Works, State Finance and Enterprise, Interior, Labor and Civil Service, Justice and Planning. The programs discussed later in this book are largely directed by women in charge of the promotion of women within the Ministries of Rural Development, Agriculture, National Education, and Interior.

Although the relatively conservative orientation of Moussa Traore's government meant intense efforts to attract private investment, Mali has attracted little and subsists mainly on foreign assistance and borrowing and its own slow development. Mali's present national debt is $822 million which is 79.4 percent of the nation's GNP (World Bank 1984, 248).

Malian Agriculture

The role of women in agriculture in Mali has been strongly influenced by the shifts in official attitude towards rural development over the last forty years. When agriculture was not the focus of attention, men as well as women suffered from the lack of understanding of—or interest in—the needs of the rural population, both as economic producers and as human beings. The particular contribution of women to agriculture was not understood because no one bothered to study it closely. Furthermore, the few programs which were started (to provide training and inputs for example), were directed at men. This was not solely a prejudice inherited from the colonialists. National planners respected traditional values and customs. According to these, women could not leave home for training seminars or take leadership roles in village gatherings where agriculture (and other economic) problems were discussed even though women might be very influential behind the scenes. Although the results of current efforts to alter this situation suggest that women could successfully have been recruited in earlier years, given the general lack of interest in traditional agriculture and the lack of understanding of the rural system of production, it is not surprising that women were neglected in agricultural programs until very recently.[2]

The current severe food crisis in Mali, however, is not due primarily to policy mistakes. Mali is one of the poorest countries in the world with a national average per capita income of $180 per year.[3] The country has been subject to creeping desertification and to the devastating droughts mentioned above. Food production in Mali has actually decreased over the last twenty years (even allowing for drought periods) (Creevey 1980), this can only be seen as catastrophic for the majority of the population, 81 percent of whom live in rural areas (World Bank 1985, 260). Most Malians survive at the bare margin of existence with no cash to buy food from outside communities. The effects of a lack of adequate nutritional foods can be seen in high rates of infant mortality, high levels of malnutrition and in low levels of energy (see Intech 1977).

Women comprise 52 percent of the Malian population and 83 percent of Malian women live in rural areas. Only 20 percent of Malian school-age girls are in school, whereas 35 percent of the boys are (World Bank 1985, 266). The proportion of rural girls in school is significantly less than the overall average. Women are counted as providing only 18 percent of the economically active rural population, but this is due to their work taking place in the subsistence sector which is not recorded in statistics. In one respect Malian women have an advantage over women in rural areas of other parts of Africa. In Mali, rural to urban migration has not assumed the proportions it has elsewhere. Only 19 percent of the total population lives in urban areas. Thus the young to middle-aged men may still be found in the family group in rural areas. Only 14 percent of the rural households are headed by women, which is likely to be almost equally the result of widowhood or divorce as rural to urban migration (Agency for International Development 1981; Republique du Mali 1981).

There are eight major ethnic groups in Mali: the Peulh (453,400),

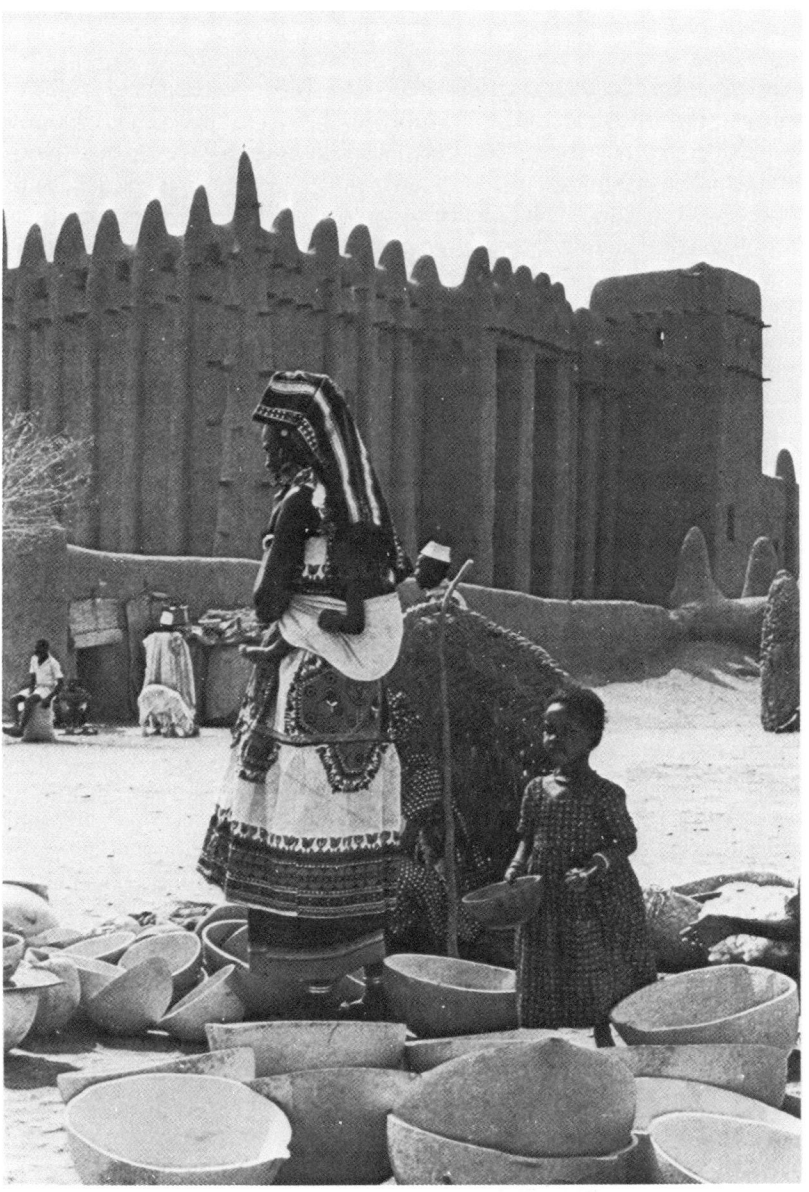

This Peulh woman with her children stands among her calabashes in the market outside the mosque of Djenne. Often these calabashes hold fresh or curdled milk and butter. Sale of such items provide Peuhl women with money for their family and personal needs, but the market is also the place of social gathering and even entertainment for women of all ethnic groups in Mali. Photo by Michel Renaudeau

Bambara (1,522,700), Sonono-Bozo (62,100), Sarakolle (446,900), Sonrai (245,600), Senoufo (319,400), Mossi (156,500), and Dogon (150,200) (Agency for International Development 1981, 2). Numerous other smaller groups also exist. In any single region several ethnic groups coexist each of whom will practice agriculture following different customs and often emphasizing different production goals. For example, nomadic groups (who are mostly concentrated in northern areas) were traditionally more concerned with their herds of cattle than with crop production. Sedentary groups, mostly in the south and central zones, grew rain-fed crops—in particular millet and sorghum—but there were variations in their methods of production. Cotton was also produced by some groups. Later corn, tobacco, and peanuts were introduced through trading and colonization. Peanuts, together with cotton, became the major cash crop for the country. Rice is also grown, particularly in the Senegal River area and along the Niger. Vegetable gardening was very rare except in the immediate area of the larger towns. During the last ten to fifteen years such crops as potatoes, yams, tomatoes, onions, carrots, green peppers, chilis, *gombo* (okra), eggplant, and cabbage began to be planted by many groups in plots wherever abundant water was available from rivers, wells, or through irrigation.

The major problem in Malian agriculture was—and is—the lack of sufficient water. Two-thirds of the land area of the country is located above the 17th parallel and is desert. From Timbuktu to Gao there is a transitional zone in which the rainfall is irregular and scarce, between 300 and 600 milliliters a year. Moving southward from Gao, the rainfall increases from 700 to 1,100 milliliters. Even in this southern zone (classified as Sudanese), the dry season lasts from October through May. The northern section of the country is not arable except through irrigation and is very sparsely populated. Eighty percent of the population is located in the Western section of the country with the heaviest density in the area between Gao and Bamako. Here rain-fed agriculture is possible, but is made more difficult because of generally poor soil often "leached by tropical rains and eroded by the wind" (Jeune Afrique 1973).

The entire problem of water is severely compounded by periodic droughts. Not only did Mali, like the rest of the Sahel, suffer catastrophic droughts in the early 1970s, but droughts have occurred frequently since then. The Malian training institute for agricultural scientists and agricultural extension workers (the Rural Polytechnic Institute of Katibougou), for example, pointed out that the average rainfall in its region from 1937–1980 was 847.7 m/m. During the last decade, however, in six out of the ten rainy seasons, the rainfall was much less than average ("Project Institut Polytechnique Rurale de Katibougou" 1982, 12).

Irrigation is practiced in various government projects (see references to a rice irrigation project in Mariam Thiam's paper following this), but is not a traditionally widespread method for coping with the low rainfall. Vegetable crops are grown on the river edge of the Niger or Senegal Rivers or next to wells, but river water is not used over a broad area for crops like millet, corn, peanuts, or sorghum. Even in recent times, efforts to introduce simple pumping devices have been stymied by the cost of pumps, the lack of spare parts and knowledge for maintenance, the extreme variance in river level in the wet and dry seasons, the shallowness of the river, and the vast expanse of mud over which the pumps must boost water.[4]

Lack of sufficient rainfall and absence of widespread irrigation become more severe limitations as the Malian populations grows (2.8 percent per year, World Bank 1985, 254). The soil has been depleted through this and through over grazing. As a result, farmers have begun to use marginal lands. Other constraints exist in agriculture as well. Virtually everything is done by man/woman power with very few tools. Most farmers do not have draft animals (oxen or donkeys). A study of sixteen villages in the Katibougou zone with a total population of 3,768, for example, showed that in 1981 there were only 198 oxen, 86 donkeys, 68 ploughs, 75 multipurpose cultivators, 41 seeders, and 81 carts available to the farmers (Project Institut Polytechnique Rurale de Katibougou 1982, 6). Most farmers do not have access to chemical fertilizers and available animal fertilizer is usually inadequate for the fragile Sahelian soil. Traditional agricultural practices are not likely to produce maximum yields. This may be because of lack of knowledge or because inputs (seeds, insecticide, and fertilizers) are either unavailable or inaccessible when needed. Or simply, as Uma Lele points out, it may be that the product of added amounts of effort is so low (Lele, 1979). Whatever the reasons, an agronomist and former director of the Katibougou Institute observed that the Bambara (men and women) farmers often do not plant seeds at the optimal time, do not space them optimally, do not add enough fertilizer, do not apply insecticide and do not weed or prune properly.[5]

Malian Women in Agriculture

One of the problems in understanding the role of women in Malian agriculture is the persistent tendency among outside observers to assume that Malian agricultural work patterns were set until the colonialists introduced cash crops and distorted agricultural practices by giving technical assistance only to men. According to this assumption, there was a fixed situation for most Malian rural women in which they were responsible for the bulk of agricultural work. They labored in the male-run family fields and had their own plots as well. They provided the bulk of food except for the meat and the grain from the family field. They made karite butter from the shea butternut tree and gathered or purchased other condiments for sauce. They sold their own surplus and decided what household necessities they would buy with it (such as clothing or condiments).

As modernization progressed, according to this theory, men received tools and advice on how to grow cash crops. Increasingly, too, men migrated to the cities. Simultaneously, birth rates rose and death rates decreased. Women, thus, had to feed more people and had more work to do (to replace migrating men) while the work they did—in the subsistence food production and processing sector—lost value relative to the higher prestige and higher productivity of the modernized cash crop sector controlled by men. Furthermore, when interest turned to food crop production, it was men who received tools and technical advice. Women, thus, may have been somewhat relieved from their labors but they lost the economic independence they had had (see Boserup 1980, McNeil 1979).

This well established stereotype of women in Malian agriculture is misleading. Malian men are indeed the major producers of cash crops but Malian women do grow cash crops as well. Malian men also participate in the growing of food crops (Barzin-Tardieu 1975). The counterview described above seems to be a result of generalizing from observations in other parts of Africa. In Mali, for example, men in many ethnic groups are dependent on the tasks women perform in their cash crop fields including planting, weeding, and harvesting. Bambara men clear the fields, young boys fertilize them with animal manure, and men build storage huts and supervise the storage of grain, except for the family surplus which women control. Bambara women receive rewards in cash for their work on men's fields. Furthermore, Bambara men help women by clearing land in their fields. Bambara women's fields are not restricted to food crops; they also grow cash crops such as peanuts on their plots and use the revenue thus obtained for family needs and personal purchases.[6]

Variations in Women's Roles

The role of women in Malian agriculture has changed continuously over time. Traditional society is neither fixed nor unresponsive to outside factors (or varying internal conditions). Long before the colonial influence, different ethnic groups migrated into the area of Mali and changed their ways of life. Some became sedentary where they had been nomadic. Some settled where land was relatively fertile and water less scarce. Some stayed for a time in drier zones which became even drier as desertification progressed. Trade routes across the Sahara introduced new tools and new ideas. As the focus of trade shifted to the coast, other influences filtered into the society. Muslim missionaries proselytized. Wars were fought to establish territory and various economic hardships such as droughts and plagues of insects were overcome. Internal organization of family and economic structures changed, though perhaps slowly, to adapt to new conditions. These changes led to considerable variation in economic practices not only among ethnic groups but also within one ethnic group depending on where the individual families were located and what their specific family situation was. As part of the overall system of life in the subsistence village, the participation of women in agriculture changed as well.

The variation in what women of different ethnic groups do in agriculture has to be considered carefully in any development program. Women in presently nomadic groups, for example, (such as Peulh, Touareg and Maure) have few agricultural tasks (although they are involved in animal husbandry). Among all groups which have been sedentary for some time (such as the Bambara, Songhay, some Peulh and Dogon) women participate in agricultural work but what they do will differ depending on their ethnic group.

Many factors will also cause different patterns even within the same ethnic group. Among some Bambara Muslims, for example, women may not work in family fields or market the family produce. This proscription, however may give them more time in their own plots. Animist women of the same ethnic group have no such proscription (Lewis 1979, chapter IX). Drought, city-ward migration and other economic hardship may change what is expected of rural women. Thus, Songhay women, who had not previously participated in agriculture, turned to working in the fields of their relatives to increase their income after the drought of the early 1970s. In so doing they increased their economic independence although simultaneously they lost status in other ways (Putnam 1978). Equally, restrictions or incentives like market accessibility and land availability will affect what women in different ethnic groups and/or different areas of Mali will do in agriculture.

Each region of Mali[7] and each administrative subdivision should be examined separately to understand the range and variety of roles played by women in agriculture. In this book, the paper by Mariam Thiam illustrates the agricultural practices of the women of Segou. The detailed task description she provides shows that, in this region, women in the sedentary groups work in subsistence and cash crop production. Their work is divided between their own fields in which food crops and/or cash crops may be grown and the family fields. They are paid for their work in family fields in gifts or in cash by the male head of the household. They market all the crops which they produce in their own fields although Mme. Thiam feels their commercial role is being undercut through modernization. Thus her discussion directly illustrates the complexity of Malian women's obligations and expectations in agriculture. As such it is a concrete refutation of the argument referred to above that women do not participate in cash crop production and do not benefit from it (in fact, they even receive wages from working on cash crop fields), although the overall impact of the process of modernization still may lead to a loss of power and independence as Mme. Thiam asserts.

Looking at the micro-level of one village or even one family helps to understand the intricacy of the gender relationship to agriculture. John Van Dusen Lewis, an anthropologist, did a close analysis of one Bambara village. His study goes beyond the scope of this discussion in its detailed consideration of family organization; however, it demonstrates that there is considerable variation in the specific economic organization of families in a single village. Household factors, such as its isolation, wealth, the power of the lineage (descent) group, the age of the woman, her marital status, the number of her children and their age, the number of co-wives, the relationship of her paternal family to that of her husband, and the religion of the family group all interact and determine what agricultural role the women plays. These factors dictate whether she has a large or small plot, whether she works in the family fields, whether she assists other women or they her, etc. (Lewis, 1979, 1981).

The distribution of agricultural tasks, including marketing, is also related to a struggle between men and women for personal and kin group power. To illustrate this, Lewis describes the factors determining whether a woman is allowed to purchase condiments for sauce. He points out that when women bought the condiments, the sauce was inevitably tastier, but where households were firmly unified, men sometimes chose to buy condiments themselves to prevent women having power over them.

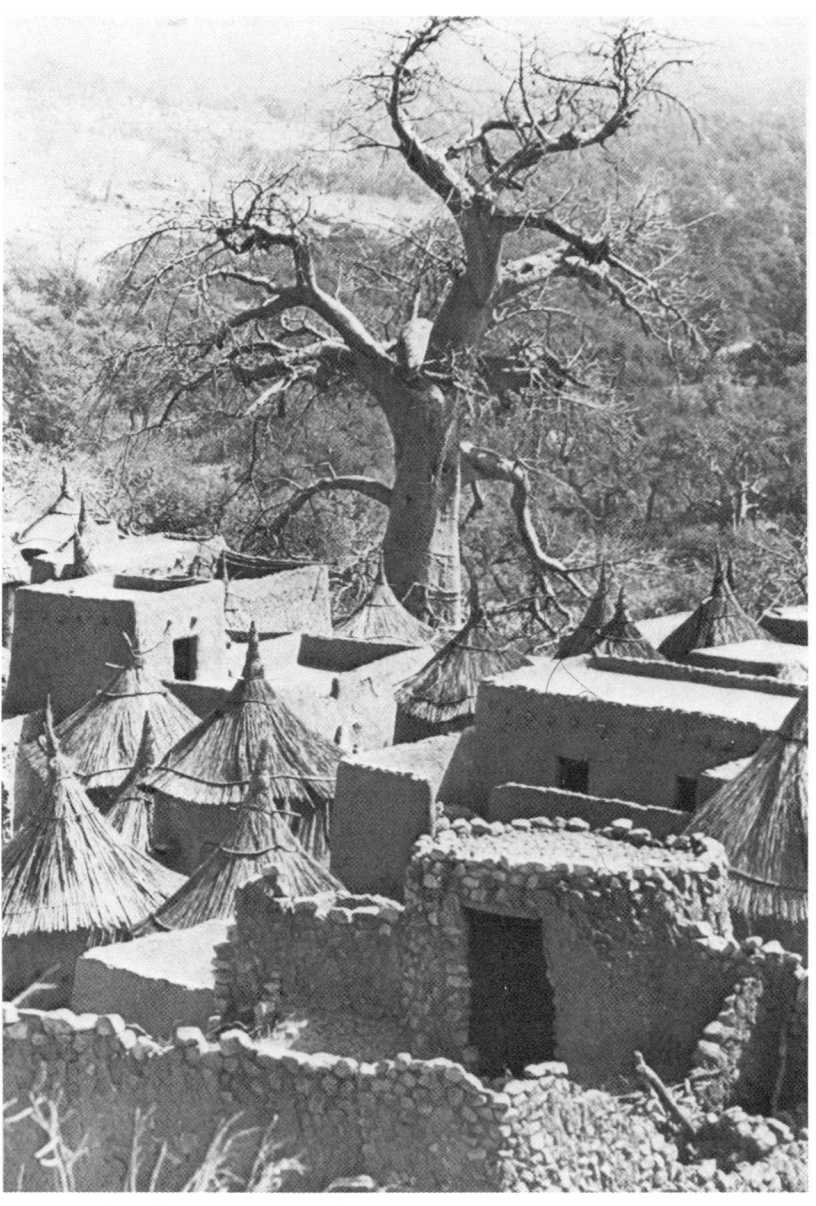

This Dogon village is composed of family dwellings called "ginna," with square terraces and granaries with pointed straw roofs for millet. Photo by Michel Renaudeau

> The lure of a tastier sauce . . . need not be added to underline the unity of the brothers, particularly if such a lure was had at the expense of investing women with some personal control, through their cooking, over the destiny of the compound. The more isolated compounds, however, had greater need of this palatal lure to sweeten the unity of its (sic) male work force. Thus . . . their women marketed more often and made greater contributions to condiment purchase than did marketing wives and mothers from the larger lineage groups. (Lewis 1979, 299)

In a summary of the economic role of women in their husband's compounds, Lewis shows that to whom alms are given, with whom a woman winnows grain and whether or not women purchase condiments is significantly related to the size of the compound and the strength of the lineage (Lewis 1979, 383).

This closer look at the micro level of a village system underlines the point that generalizations about the work of all Malian women may be misleading and are even misleading in the context of a single ethnic group. Plans based on this type of oversimplification are bound to fail. Planners will not understand why some women embrace a new program and then abandon it, while others refuse it at first and later come to accept it and a middle group gets only partially involved. The only possible approach is an individual/family/village strategy which emphasizes local participation—local goal setting and local evaluation as projects progress. This type of planning must be flexible, allowing for changes as circumstances are altered. Women could thus be reached according to their particular situation at a given time in a given place.

Factors of Production

One set of questions is essential to planners seeking to work with Malian women; these relate to the ownership of the factors of production. The complexity of this subject is suggested by the preceding discussion of the great variety in economic practices and some reasons for this variety. Traditionally, land was not owned by individuals but by families; the allocation of that land was by the family head, a male. Ownership, however, is a misleading term in this context. In many Malian ethnic groups land was used as a kind of feudal grant by groups of families constituting, in scattered compounds, the villages of Mali. There was no real ownership, but use after an original grant and long custom. Sometimes groups of strangers for some reason were given land. Land is not scarce in Mali (although good land is), thus Mali has not had the same pressure on this resource as have other parts of Africa. But the question of proving individual ownership of land, inevitably arising in the modernization process, is nonetheless complex. Most Malians still follow the traditional practices without legalizing their position. In this system, the position of women generally is determined by the head of their household. Of course his decisions are determined by many factors other than his good judgment or his preferences, such as custom, the amount of land available, the number of grown sons or wives, etc. As a broad generalization, however, whether a woman has a plot of her own is decided by her husband or father. Where a family does establish legal rights to the land, this will be done by the family head who will allocate its use thereafter and who is almost always a man.

The same problem exists with any of the more complex means of production in agriculture. Observers have long pointed out that projects to develop agriculture have given all inputs, training, and equipment to the men. If they had not done so and equipment was given to a family unit, the same general result could have been expected. In the existing social system, control of the draft animals, carts, ploughs, etc. is generally a part of the male head's economic direction of his overall family unit. Giving tools to the women alone would have been unacceptable in the traditional system and still would be difficult in the present society.

A Malian woman has her own labor but even the use of this, by custom, is determined by many complex family needs. She often has the right to determine the use of whatever surplus profit she makes in her own fields (even though she may use this for household needs). But the amount of time she has to produce that surplus, and whether she has the land or the right to market, is often at least partly outside her control. Men, too, are forced by factors outside their control to work in certain ways and to fulfill certain obligations, but women were traditionally and still are subject to men in the economic decision-making hierarchy of the family unit.

It would be wrong to assert that Malian women have no power or status in the agricultural process; they do within most ethnic groups and in most circumstances. But women are at an obvious disadvantage in Mali. In the process of modernization, their independence on male decision makers within their own households and in local, regional, and national government positions has led to their own interests as a group being ignored or given second place. Sometimes women do not get land in a project and they may boycott (see Dr. Cloud's paper, page 43). Sometimes they get less land or the worst land and can do nothing. Sometimes they get no tools (as these are given to men farmers) or the machines will be used by men on women's land, or in family fields to replace traditional tasks performed by women for money or gifts, or even power. Bernhard Venema's discussion (the last paper in this section) gives an excellent example among the Wolof of complex changes in social relationships occurring because of crop mechanization. As he shows, some of the changes were not foreseen and certainly not intended by the policy makers. Many changes were directly detrimental to rural women—increasing their work load while reducing their power and authority within their local society. Unless development projects are drawn up taking into account the impacts of any proposed changes on the position of women— even if they are not the prime targets of the project—women will lose further ground with a consequent heavy social cost both for women as a group and for the production of food to which they are so essential.

Conclusion

Women contribute significantly to agriculture in Mali. They provide a substantial part of the labor used for the production of both cash and food crops, and they play an important role in the care of animals and poultry. Because of the way society is structured in rural areas, women still work in what are seen as women's tasks and are regulated by a special set of rules pertaining to them only (modified to suit their particular family status). It is necessary to consider them apart from men in analysing the problems of and prospects for rural development. On the one hand, in order to reach women farmers any successful project will have to look at what men do and how they react. On the other hand, if women are not singled out to be directly assisted in learning new techniques or receiving tools and equipment, they will lag far behind men in changing their habits and outlook.

In Mali, society was more reluctant to allow young girls to be educated than young men. Thus a far larger proportion of rural women than men are illiterate. In addition, far fewer of them have had the experience of going to a town for a short term job with all the training and orientation which that implies. It is not surprising that rural development projects for women have often found their eager participants unable to carry through the tasks they took on. Simple things like keeping track of what they put into a project and how much they get out, following instructions for the application of water and fertilizer, or taking stock in commercial ventures are very difficult without experience or ability to read or write, add, or subtract.

The general situation of women in the agricultural sector in Mali is the same as that of men. For both, life is very hard. Lack of water and sufficient resources to compensate (by drilling for wells) is the single most severe problem. Government policies, which encouraged the growth of cash crops and did not provide adequate incentives to make food crop production equally attractive, have distorted the pattern of agriculture. Lack of credit for poor farmers, poor transportation facilities and road networks for marketing, unreliable timing of deliveries of inputs such as seeds and fertilizers (chemical), nonavailability of simple improved tools, and the costliness of draft animals all contribute to the plight of the farmer.

The government of Mali has sponsored many programs aimed at alleviating this situation. Credit programs, marketing cooperatives, improvement of roads, training programs, functional literacy programs, and experimentation with improved seeds and different kinds of fertilizers are only a few of the things which the government has undertaken.

Despite the lack of resources and the seriousness of the agricultural problems in Mali, the government has managed to develop programs specifically for rural women. The government has become increasingly aware that without these the overall agricultural situation will not improve. These projects have had some encouraging, even exciting, results (see papers by Kantara, Djire, Traore and Diallo, Section II). But change only comes about gradually and there is still considerable debate as to how to handle the many issues and questions which arise as the various programs develop. In the meantime, the overwhelming majority of rural women in Mali follow traditional rules and practices and are largely untouched by the efforts which have been begun for their benefit.

Notes

1. Aly Cisse is former minister of Public Health and Social Affairs in Mali, former executive secretary of CILSS, and current executive secretary of the Comité International du Corps pour L'Alimentation (CILCA).
2. See discussions of Malian political history and social and economic development in Morgenthau and Creevey 1984, Snyder 1965, Ernst 1976, Jones 1976, Jouvre 1974, Mega-

hed 1970, Hopkins 1972. For discussions of Malian rural women see the bibliography produced by Susan Caughman (Caughman 1984).

3. Mali is surrounded by Mauritania, Senegal, Guinea, the Ivory Coast, Upper Volta, Niger and Algeria. Only 27 percent of primary school-age Malians are in school. Life expectancy is 45 years and the average index of food production per capita is 83 in 1980–81 (1969–71 = 100) (World Bank 1985, 218, 228, 266).

4. Fact-finding report on the Katibougou Project presented to the Comité International de Liaison du Corps pour l'Alimentation (CILCA) by Rob Hecht and Samou Sangare, March 1982.

5. Dr. Samou Sangare, former director of Katibougou and regional CILCA coordinator located at CILSS Ouagadougou 1978–1982, in a planning session, December 1982.

6. This generalization is true for most Bambara farmers in Mali but not all. Thus Bambara Muslims have a different allocation of labor as mentioned in the text above.

7. There are seven economic regions in Mali: Gao, Kayes, Koulikoro, Mopti, Segou, Sikasso, and Timbuctou. The area around the capital city of Bamako is separately designated as a "district."

3 The Role of Women in Rural Development in the Segou Region of Mali[1]

Mariam Thiam

Editor's Note

Mme. Mariam Thiam is a member of the National Women's Union of Mali (UNFM). She has worked on many women's projects and is an active and vocal contributor at sessions where the issues of how to work with rural women in Mali are discussed.

This chapter was presented to the American Friends Service Committee with which she collaborated for many years on a project to increase income opportunities for women. Although it was written in 1976, Mme. Thiam's observations on the Segou region are still considered valid by those engaged in projects in that zone. The chapter is of particular interest because it illustrates in considerable detail what women of different ethnic groups do in one region of Mali and what one Malian woman, directly involved in development efforts, thinks are the major problems facing rural women.

The chapter cites five development projects in Segou: The Niger Bureau, The Segou Rice Project (ORS), the Peanut and Food Staples Project (OACU), the Malian Company for Textile Development (CMDT) and a Fishing Project. It also mentions the Centers of Training for Rural Development (CAR) and the National Headquarters for Rural Development and Training attached to the Ministry of Rural Development. To explain these references, the government of Mali embarked in 1972 on five major national development operations aimed at improving the rural economy as a whole. These included the Rice Project and the Peanut and Food Staples Project mentioned by Mme. Thiam as well as an Animal Husbandry Program and a Cotton Project. Planning for these projects included vast training programs and the provision of inputs and equipment to farmers. Donors were attracted from many different countries. At the outset, women were not included among the farmers to be trained (see Ly 1975) but considered in support activities to improve health and general welfare.

Training for farmers in these programs comes from Functional Literary Centers (see Djire's paper in Section II) and from Centers of Rural Animation (CAR), and through the direct action of extension agents visiting individual villages and attending village meetings. The CAR originally took groups of men farmers, chosen by the extension agent and village leaders, who were to return to their villages to spread the new techniques and approaches they learned (in courses from a few days to a few weeks in duration). From the late 1970s onwards women also attended these sessions.

The Office du Niger (Niger Bureau) from colonial times was charged with developing programs and projects for the Niger River area. In colonial times these included rice and cotton projects and various fishing programs. More recently it has supervised many rural development projects including those led by Mme. Thiam in Segou.

Introduction

Since Independence in 1960, the Republic of Mali has shown great concern for developing both its economic and human resources. This is understandable particularly since during the colonial period the country, then called the French Soudan, was ranked among the least favored, least "interesting" of the French territories because of its enclosed geographic position, its distance from the sea and its situation almost entirely in the Sahel. There was little investment in development or in socio-economic infrastructures in the entire area, and even less in the rural sector.

Spread out over a surface of 1,240,000 square kilometers, Mali today (in 1976) has a population of 6,300,000 inhabitants, or an average density of 5.08 inhabitants per square kilometer. The majority of this population (about 85 percent) live in rural areas where the principal economic activities are agriculture and cattle raising. In spite of the long period of drought, a calamity that peaked in 1973–1974, the level of participation of the rural population in domestic production is very high. Its contribution grew from 58.7 billion Malian francs in 1969 to 110.3 billion in 1978.[2]

To facilitate the administration of this vast territory, Mali was divided into regions, *cercles* (sub-regional administrative units), *arrondissements* (sub-*cercle* administrative units), and villages. These are structures based as much on the needs for administrative decentralization as on the socioeconomic environment. Mali has seven economic regions and one district (Bamako), 46 *cercles,* 281 *arrondissements* and 11,000 villages.

The Segou Region

Segou, the fourth economic region of Mali, is situated between 13 degrees and 16 degrees latitude and 4 degrees and 7 degrees longitude. It is the smallest Malian region, covering an area of 62,889 square kilometers or 5.2 percent of the national territory. Segou is bordered on the northeast by the Region of Mopti, on the south by the Sikasso Region, on the west by the Koulikoro Region, Mauritania in the north, and Upper Volta in the southeast. It is relatively flat, made up of plains on each side of the Niger River.

Climatic variations within the region are determined by its North-South configuration. Indeed, the climate is wetter in the south, called Soudanese, becoming Sahelian and sub-desertlike in the north. The Segou Region receives its water supply from two rivers which are essential for its economic life: the Niger and the Bani. The Niger River crosses the region for 292 km of its 1600 km course through Mali. It also supplies an irrigation network through the Markala Dam. This has made Segou a model agricultural region.

Segou's essentially rural population works in agriculture, cattle raising, and fishing. There are 1,068,000 inhabitants, composed mainly of Bambaras, Bobos, Somonos, Bozos, Peulhs, and Miniankas. The Bambaras constitute the largest ethnic group.

Except for the Peulhs (cattle raisers) and the Bozos (fishermen living along the Niger-Bani Rivers), all the other ethnic groups are involved in agriculture. Nearly 90 percent of the working population is in the agricultural sector, and there is an overlap and complementarity of cattle raising and agriculture. Aside from agriculture, cattle raising, and fishing, the population practices food gathering and vegetable gardening, thus obtaining a large part of its revenue.

The population density of the area has the greatest concentration at Segou ($35.4/km^2$) and the sparsest population at San ($27.4/km^2$). Only the north has a population density below the national mean, with four inhabitants per km².

The inhabitants of Segou are almost entirely Muslim, but they nevertheless retain fetishistic customs and still hold traditional ceremonies based on the cult of the ancestors, consultation of oracles, and sacrifices to the gods. Some have embraced Christianity, but they have not abandoned, to any great extent, their ancient customs.

Considering the economic potential of the Region and its human resources, several projects in socioeconomic development have been undertaken at the initiative of the government. The Region has five of the most important development projects of the country (see Editor's Note):

1. *The Office du Niger (Niger Bureau)*, which was founded in 1943 and became a state enterprise in 1962, deals specifically with the cultivation and improvement of the Niger Valley. With its dams, canals for irrigation, navigation, and water supply, it produces rice (100,000 tons per paddy per year) and sugar (20,000 tons of powdered sugar per year). It has an agricultural-industrial and commercial character which benefits the peasants (called "colonizers" in the area).

2. *Segou Rice Project* (ORS), implemented in 1972 with the technical and financial assistance of the European Funds for Development (FED), cultivates almost 40,000 hectares of rice paddies divided into fourteen sections in the middle valley of the Niger. The workers are organized not only to increase their production, but also to improve the socio-cultural conditions of their lives. Five community development centers are in operation there in order to solve problems of hygiene, nutrition, maternal and child health care, and literacy among participating families.

3. *Peanut and Food Staples Project* (OACV). With the assistance of the World Bank, this project works to increase the revenue of peasants through the introduction of selected grains and new methods of cultivation. Its social branch plans health, literacy, and education activities for the peasants.

4. *The Malian Company for Textile Development* (CMDT) deals with the development of cotton-growing in Mali as well as the cultivation of *dah* (a variety of the sisal plant). The production of *dah* falls within the scope of integrated projects because the harvested crop is almost totally processed on the spot by the Malian Textile Company at the textile plant at San.

5. *The Fishing Project,* which benefits from the assistance of FED and its office in Dioro, organizes the fishermen of the Niger and Bani River Valleys and helps with fishing techniques as well as marketing the fish.

Along with these development projects, Centers of Training for Rural Development and continuing education classes are held under the auspices of the National Headquarters for Rural Development and Training which is attached to the Ministry of Rural Development. Some of these CAR centers are coeducational, allowing young rural couples to complete their training together. This is the case of the C.A.R. in Yangosso.

As we see it, the existence of these numerous development projects is a major advantage for the improvement of productivity and of the quality of life of many peasants in the Segou Region. Within this context of well-being applied to the entire community, it seems that one of the essential conditions to be met is the emancipation of the woman and her participation in development. Let us look at the situation for women nowadays in this Region.

The Specific Role of Women in the Development of the Region

In asking ourselves what rural woman do in the Segou Region, it seems necessary to distinguish among different levels of participation of women. Let us examine first the case of women who are not engaged in any agricultural activity. This case is relatively rare in this region, but it is found in several villages or in large families which are very devoutly Muslim, for example, in the regions of Sansanding, Dioro, and Baroueli. These women (in purdah) do not communicate with the outside world except through intermediaries—their husbands or their children. Their sole activity in addition to housework is in handicrafts, sold by their intermediaries.

Except for this tiny group of women, which one would consider neither liberated nor protected, we find all the others in family-based farming, for the Segou Region has the reputation for being the granary of Mali and is involved first and foremost in agriculture. This activity is not considered to be the job of an unmarried person; it is the livelihood of a couple or of a group, where men and women have complementary tasks to fulfill and where it is difficult for some to get along without others. This complementarity is a cohesive element for married couples.

To describe the activities of women in the rural world, we have found it interesting to draw up a comparative list by ethnic group of their

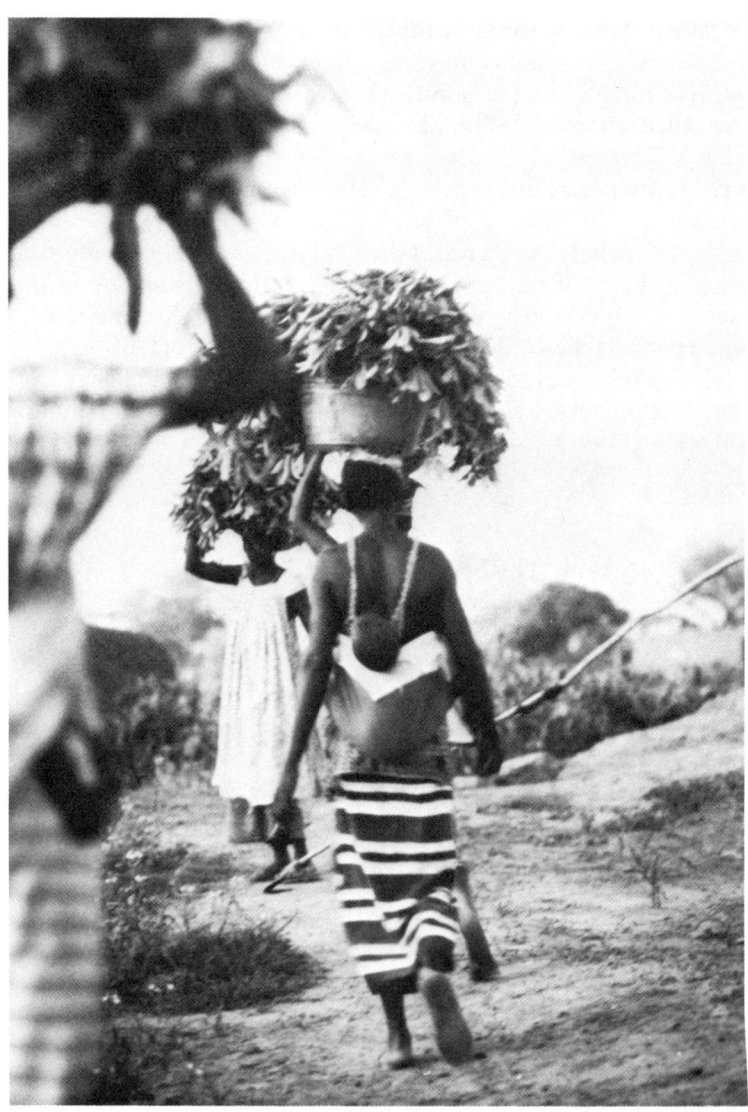

One of the many arduous tasks of Sahelian women, illustrated by these Bambara women in Mali, is searching for and carrying home firewood. As the population expands and the plant cover is depleted, the task becomes more difficult. Deforestation threatens Mali and other Sahelian countries with erosion and the loss of precious soil as whole areas are turned into dust bowls. Photo by Richard Harley work in the traditional milieu and then do the same in the context of "development projects."

The Work of Women in the Traditional Sector

1. *The purely agricultural sector.* The following example describes activities of the Bambara women in the villages of the Segou circle (village of Sakoibougouni situated 30 km from Segou).

Food production. In the family-worked fields, they participate in all tasks: transportation, spreading organic fertilizers, weeding, banking, securing the fields against predators, harvesting, and transportation of harvested crops.

On individual plots of land, granted to them by their husbands (i.e., the land belonging to the household), they work the fields early in the morning when they are not cooking, before beginning to work on the collective field, or when the sun is high and when they have ceased working on the collective field in order to rest. On these plots, they are obliged to clear the land and to work hard if they have no sons or sons-in-law to do it for them.

Gathering. The women gather leaves and fruits (sorrel, *gombo* (okra), *dah* fibres, bean leaves, baobab leaves, karite nuts, wild grapes, vines, monkey bread, *nere,* etc.)

Processing. They process the agricultural products and the plants they have gathered into beer *(dolo),* doughnuts, flatcakes, dried gombo, *soumbala* (a condiment which is used as a powder in sauces, and is similar to curry), karite butter, and native soap.

Cotton processing. They spin cotton to make clothing for the entire family.

Market gardening. They farm small kitchen gardens which produce onions, tomatoes, sweet potatoes, manioc, pepper, beans, gombo, and other condiments. At Bambougou, they also farm tobacco along the riverbank.

Livestock. The women raise poultry and small grazing animals always for the purpose of meeting family expenses if the need arises.

Crafts. The wives of blacksmiths are potters. Natural and indigo dyeing processes are practiced by low castes. Basket weaving is a feminine art, whereas cloth weaving, tool making, sewing, and embroidery are done by men.

Commerce. The women work in small-scale commercial activities: the sale of surplus food (grain, processed food, fresh garden products, drinks, and cotton goods—from cotton which they have spun themselves).

Remarks. This general framework applies to the Bobo women in the sectors of San and Tominian as well. In addition, some Bobo women specialize in pig raising, whereas the Minianka women specialize in preparing Soumbala, making karite butter, growing red pepper, and making calabashes.

2. *The fishing industry.* The fishermen of the fourth Region are Somonos. Somono women are engaged in the following activities:

Household tasks. Cooking, dishwashing, washing clothes, grinding millet, purchasing wood (or searching for it in the bush), supplying water, raising poultry.

Agricultural tasks. At Kirango Somonobougou, they raise onions and potatoes along the river.

Fishing. Only young girls help to drag the large nets. They share in the distribution of the catch. It is not rare to see women fishing in lakes or ponds.

Fish preparation. Fish processing such as drying, smoking, roasting, and grilling are done by the women.

Marketing. They are intrepid vendors. They sell fresh fish in small quantities at the local markets. They are wholesalers and retailers of smoked and dried fish. Some rich women traders are owners of fishing canoes which they rent to young fishermen during the week. They are also sellers and "resellers" of condiments imported from other regions and of rice.

3. *The cattle-raising sector.* The cattle raisers (Peulhs and others) of the fourth Region are confined to the area of Macina and are scattered among the peasants in the other districts. They are semi-nomadic or sedentary. In their villages, the wife gardens in the field adjoining her hut in addition to her household tasks. She processes and sells the milk which she carries on her head, transporting it long journeys by foot. She raises small animals (goats and sheep) and particularly poultry (the sale of guinea hen eggs is a well-known activity of these women). In the Macina *cercle* women weave baskets and mats.

The Work of Women in the Cash Crop Sector

The need to export increases with the need for tools and equipment in the country and explains the importance attributed to cash crops such as cotton, *dah,* peanuts and rice. The women participate in all the work of raising these crops in addition to providing food for the family. They are rewarded either with gifts or are paid farmhand wages. They always tend their individual plots and gather and process plant products as well as sell what they produce.

The managers of large development operations in the region are aware of the importance of women in this agrarian export economy and have included the services of "relief and progress" in their programs [especially to help women]. For example, the community development center included in the Segou Rice Project operates in four localities of the Region taking charge of programs in nutrition and sanitation education and family economics, etc. These community centers take account of the economic role of rural women, and have other training services as well to improve techniques in poultry raising, product processing, handicrafts (particularly dyeing processes) and gardening. Family revenue has grown because of this. The Community Development Center organizers (CDC) have tried to lighten the work load by introducing labor-saving machines (plows, grain mills, etc.). Using animals to plow by yoke or harness spares the arduous work of the *daba* (hoe) but calls for a large investment which must be paid for. It allows one to farm larger areas but these must first be seeded by rows and then harvested. On the whole, production is perhaps greater, more money is earned (loans for the plow and the fertilizer must be reimbursed) but the woman does not work less.[3] It would be interesting to do a more careful study of the impact of harness plowing on the participation of women in the fields (see Venema below for discussion of the impact of oxen among the Wolof).

The use of the grain mill surely presents many advantages. It is used more frequently for grinding millet, beans, karite nuts, and for preparing *dolo*, butter, flat cakes, and donuts for the market. Time is saved, fatigue is spared, and it is more profitable. But one must consider cost; the mills generally belong to businessmen who profit from an almost complete monopoly in the village and set the prices at their own convenience.

Some women in the Region use cotton gins introduced by the CAR, at the coeducational training center at Yangasso, for example.

The dilemma is the same: time is saved but money is needed to purchase machines.

Remarks. Thus, access to new techniques is expensive and is becoming all the more so, particularly when imported material is involved. It necessitates an increase in income generating economic activities. Given the actual conditions of the division of family revenue, this generally means more work, particularly for the women.

The Women's Situation

Concluding this comparative study of the activities of women, we are able to note certain changes in the rural milieu in the Segou Region. Both in the traditional sector and in the sector of commercial farming, modernization is occurring which has positive and negative repercussions. Until independence, women controlled interior commerce almost entirely by themselves. They frequented the markets located sometimes quite far from their villages in order to sell condiments, grains, beverages, etc. Thanks to this commerce, they had undeniable financial power. In any case, the husband is supposed to supply only a house and grain to his wife by custom in Bambara society; the rest is the personal responsibility of the wife.

But after independence,[4] the country developed cash crop farming (peanuts, cotton, rice, dah) and the commercialization of these products became the prerogative of the man (in spite of the additional work that these crops entail for the woman).

When women have succeeded in developing a business and this business enterprise reaches a certain size, the management of the enterprise now passes into the hands of the men. This is the case, for example, in the cultivation of onions in Bobo territory (the *cercles* of San and Tominian), and in the sale of fish to the large cities of the region by the Somonos of Markala. Men also sell wood and water in order to pay for their carts, although often their own wives continue to carry wood and water on their heads.

Thus, with the appearance of men in the marketplace, we pass from an economy based on self-sufficiency to a monetary economy in which women progressively lose revenue sources as men further monopolize the markets. At the same time, their dependency vis-à-vis men is growing (*Editor:* see discussion in Venema below).

There are counter-trends to this general situation which include the following examples: To organize the varied tasks involved in upkeep of the home, work in the fields and other activities such as fishing, cattle raising, and marketing, the peasant women of the fourth region have established a separate program for themselves which varies according to village and ethnic group. This separate organization most often makes the woman financially independent of her husband [*Editor:* as she manages her own affairs apart from his]. For example, it is a grave offense for a Bambara woman of Segou to hear her husband say in public that he buys her clothes for her. A popular saying of this group is: It is for the husband to supply the *to* (millet dough which is at the base of Bambara cuisine) and for the woman to supply the sauce.

In this region the customs of different ethnic groups stipulate that the land belongs to a lineage, which includes ancestors as well as the descendants not yet born or those to be married. That is why fields are attributed to women, either individually for home gardening or collectively. [*Editor:* rather than being assigned to her by her husband or father at his whim]. This is the case, for example, in Minianka territory where the wife is the exclusive proprietor and she disposes of her field as she sees fit. A second case is that of the wives of migrant workers, men who leave home seeking their fortune (in Ivory Coast or in Bamako). They leave their wives behind responsible for farming, for the elderly, and for the children. These women become farm managers without having asked for it, simply by the force of circumstance. The wives of cattle raisers own animals and place their husbands or a nomadic shepherd in charge of them. This is a form of savings and of securing capital for them, and is also an element of prestige (the wife's dowry often includes several head of cattle). Somono women can be canoe owners or have a large sum of capital from their commercial transactions.

Finally, the rural women of this region meet their own expenses in different ways, according to the generation to which they belong. The oldest do so by means of their intense activity in agriculture, cattle raising and small commerce. The youngest, particularly the young girls, do so by working in the cities during the off season just as the young men do.

On the social plane, women participate in festivities, baptisms, marriages, and funerals. They can be members of various associations and groups, and even of secret societies in which, as "witch doctors," they have great influence on their fellow citizens because they are feared.

From what has been previously stated, it is apparent that the increased development of cash crops has upset the balance of the traditional self-sufficient economy and the entire economic, social, and cultural system which depended upon it. The new system is based particularly upon a growing monetarization (in order to satisfy the new needs). The stereotype of the rural woman as a person devoted to social functions while the man is the economic agent has led to an impasse. At the present time, we must choose modern techniques and introduce them in the interest of the rural society as a whole for both men and women. These techniques must be within the capacity of local artisans who will assure the maintenance of the materials and equipment.

The search for this new balance is based upon principles such as the following: (1) Choose simple machines which, if not manufactured in rural artisanal centers, are at least maintained and repaired there, and which will alleviate the burden of men and women in their agricultural, artisanal, and processing activities; (2) Organize men and women in village or inter-village groups in programs intended to mobilize capital to expand production while alleviating their work load through use of collective equipment (wells improved with pumps, grain mills, plows or tractors, etc.) and improving the conditions of village life (through sanitation, care for the environment, health and cultural facilities etc.). These are preliminary requirements for a new economic order in the rural area. If these ideas were adopted by the government, they would spread rapidly. For the peasant, whose wealth depends in large measure upon the number of hands working in his field, there can be no other wealth than that of human labor, as long as no machine, no matter how rudimentary, is within his reach.

Once liberated from the constant struggle for survival and the satisfaction of primary needs, the rural world will turn resolutely towards the search for higher and more sophisticated leisure activities in the socio-cultural domain. It is then that rural people will be able to comprehend the notions of individual well-being (including concepts of family planning) which they will then wish to apply in order to effectively improve their own situation in the Sahelian zones disfavored by climate and ecology. Then, for instance, women will go to the PMI centers for consultations and social services, maternity and family planning care.

Conclusion

The role of women in rural development in the Segou Region, as we have just seen, is certainly very important. On all levels and for all ethnic groups, women contribute to the process of producing, processing, marketing, and distribution of goods. They must consequently receive education and training. Within this framework, rural development programs must give as much importance to women as to men, and not tend, as is now the case, to give priority to men with the excuse that programs thus move more quickly. If the rural woman does not participate in and benefit from all innovations, many sectors of society, such as the education of children (particularly pre-school) and the proper use of revenues for healthy food and good hygiene will remain backward. Rural development planning must include programs which will encourage and facilitate the integration of women into the modern sectors of the economy.

Notes

1. This chapter was originally written in French and was translated by Mimi Mortimer.
2. 410 Malian francs = $1.00.
3. *Editor's Note.* The problems here are in regard to purchasing the equipment and the actual impact of these tools on women's work and production. Rural women do not have the cash to pay for these tools. If they are given on credit, then the women have a heavy debt which they may not be able to pay if any negative factor intervenes. Furthermore, the effect of the tool may not be labor saving as intended. For example plows allow a greater area of cultivation but this means more work in the stages of planting, weeding, and harvesting (see discussion of these constraints in Lele 1975).
4. *Editor's Note.* Cash crop farming began in the French colonial era, not after independence. Mme. Thiam means that it was intensified after independence.

4 The Changing Role of Women in Sahelian Agriculture

Bernhard Venema

Editor's Note

Dr. Bernard Venema is lecturer in sociology and anthropology of North and West Africa at the Free University in Amsterdam. His chapter here concerns the Wolof of Senegal, but what he shows about the complex power and dependency relationships between the sexes and the way these change due to outside influence including the introduction of cash crops and of modern tools and machinery is relevant for other Sahelian societies. Dr. Venema refutes the notion that women do not profit from cash crops or have any benefit from the introduction of modern tools. But his paper clearly demonstrates unconsidered side effects from mechanization programs and other "modern" changes in terms of women's work, their general obligations and their status. These side effects can be positive or negative— Dr. Venema illustrates both—but, he points out, the lack of understanding of the impact of modernization on a complex social system as among the Wolof in general may lead to inadequate planning with sometimes unintended severe hardship for some family members and particularly women.

Introduction

Early in 1973, members of the Senegalese research institute, the Institut Sénégalais de Recherche Agricole (ISRA), organized a meeting with farmers of Sonkorong [an area in the Sine-Saloum Region] to discuss a more efficient use of agricultural implements. They told the farmers they were neither interested in a reduction of the number of personal plots in the farm-holdings nor in how fields were distributed. But they argued that labor and farm implements would be better employed if all farm laborers (and equipment) worked on each plot in sequence, the exact order to be decided by the members of the household. It appeared, however, that the women did not agree with this proposal because they felt it would arouse feelings of jealousy among co-wives who did not have plots of the same size. When it was suggested that all women should have a field of equal size, the farmers stated they would have problems with their parents-in-law, who would argue that their daughter was able to till a large plot. Among male adolescents, the general preference was to work on their own plot when and with whom they wanted, and they were afraid that a change in the distribution of labor and equipment would be even more favorable to the heads of household. The ideas of ISRA were not put into practice.

In spring 1971, the same farmers were informed that in compensation for the crop failure of the preceding year every farmer would receive 1900 CFA francs.[1] This money came from the EEC and it was stipulated that it was destined for every millet and groundnut grower. When the farmers obtained the money, they kept it for themselves. The adolescents in the village became very angry. They claimed part of the money because they also had cultivated groundnuts. In addition, they had assisted the heads of household in millet and groundnut cultivation on the communal fields. To voice their protest, a number of male dependents remained at home several times on mornings they were required to cultivate the communal fields. However, they did not succeed; the heads of the households still kept the money. They promised, however to use all the money to buy millet in the 'soudure' (hungry season) (Venema 1978, p. 115).

These examples show that the Wolof farm is not a unified production unit, but consists of a number of separately cultivated fields. Consistent with this, there is no central management of the farms. Every adult member of the household manages his own plot. Autonomy in the management of the personal plot, however, is only relative because the usufruct of the plots is founded on a system of reciprocal rights and duties tying the individual holdings together. So, in fact, the individual plots constitute one farm holding.

The several rights and duties in Wolof farm-holding have a long tradition. Anyone who tries to change them will meet with difficulties, as was shown in the preceding examples. Francis Moore, a factor [company agent] working on the River Gambia in the eighteenth century, drew attention to what was clearly a well-established division of labor between the Mandinka men and women in Gambia. He observed that, "The Men work the Corn Ground, and the Women and Girls the Rice Ground . . . and the Women are busy in cutting their Rice; which I must Remark is their own Property; for, after they have set by a sufficient Quantity for Family Use, they sell the Remainder, and take the Money themselves, the Husbands not interfering" (Moore in Dey 1981)

In regard to the Wolof, Ames (1953) and Monteil (1967) observed that at least from the eighteenth century onwards, unmarried male dependents cultivated farms of their own. Normally they worked five mornings of the week for the head of the household and used the evenings and whole days of Thursday and Friday for cultivating personal plots. Domestic slaves also followed this system. This practice continues today.[2]

Because production in a Wolof farm results from a balanced system of reciprocal rights and obligations, one could draw the conclusion that no changes in the way the farm is managed are possible. However, a new balance in the system is possible as a result of changes in the environment outside the household. To explain shifts in the relative positions in the household, it seems useful to consider each member of the household as a person trying to maximize profit from externally induced change. The before-mentioned aid given to the farmers showed that EEC and Senegalese officials believed the head of the household represented the farm. Because of this misunderstanding, the heads of household were able to pocket the money intended for everyone. Once this had occurred, the male dependents could do little to correct things. They wrote to a popular radio program in order to get its support, but the procedure lasted too long and they had to give up. Some decades before the male dependents had been more successful in improving their position.[3]

Most relationships, especially in tribal and peasant societies, are complex. It would be a mistake to see relationships between people only as determined by material considerations for they also are governed by social and religious factors. A Wolof woman, for example, cannot do all the agricultural work on her plot because of the sex-division of tasks. From this it could be deduced that she is completely dependent on her husband to have the male tasks on her plot performed. This is not so because she uses her position as mother, mother-in-law, 'godmother' ('dieukee') to obtain male labor.

Although material considerations can not explain all behavior, ample attention must be given to them in order to understand African agricultural systems. If we confine our analysis to the man-wife relationship, it would appear that many monogamous households have a common budget with man and wife jointly tilling the farms of the household. In polygamous households, the relation between a man and his wives, however, is based on individual interest as is shown by the cultivation of a personal plot by the wives and an exact division of responsibilities for household expenses. This is due to the strict demarcation of rights and obligations in polygamous households in order to diminish rivalry between the cowives. If the man is not able to hide the fact that he has a favorite, this may result in heavy strains within the household. Although most men only have one wife, polygamous marriage may serve as a general model for the position of women because most men believe in [the propriety of] polygamous arrangements. The joint collective tilling of the fields of the farm holding is an exception and this type of collaboration was found mainly among recently married couples.

Difficulties in marriage relations are one reason a Wolof woman maintains good relationships with her own family. She maintains contact with her parents, her mother's brothers, and her own brothers by visiting them regularly and it is these people to whom she turns in case of troubles, divorce, or widowhood. Good relationships with the maternal kin also result from the fact that witchcraft, luck ("baraka"), and diseases such as madness and leprosy are thought to be transmitted through the female line. Therefore, for trust and love, a woman turns to her maternal kin (Venema 1978:98–101). For travelling expenses and gifts, a woman needs an independent source of income to maintain contact with her natal village. Thus ties with the maternal family and the role of polygamy help to explain why Wolof women hold on to their economic independence. Similar arguments are used to understand [women's] economic independence in other societies in the Sudano-Sahelian area such as the Bambara or Haussa, and Fulani (Belloncle 1980; Dupire 1960; Caughman 1981).

In implementing agricultural policy, one cannot say, therefore, that the head of the household necessarily represents the interests of the household. Neither can it be argued that there are independent economic spheres for man and wife, allowing for the introduction of change for one without affecting the position of the opposite sex. Because farm holding is a system, innovations will have effects on all members of the household and the end result will often be the continuation of the balance of power within the household rather than a radical change.

The Introduction of Cashcropping and its Consequences

According to Boserup (1970) and Goody (1977) the introduction of cashcrops in subsistence agriculture increased the role of men in the total work input in agricultural production. In an agriculture dominated by cashcrops there develops, they argue, a male cashcrop system parallel to a female food crop system with the men receiving the cash income and the women cultivating food crops and vegetables as food supply for the household. With these she supports herself and her children: the surplus, if any, is sold for personal expenses. This view is also taken by many donor organizations including the Dutch government (MIN 1983: 101).

This radical shift in sex roles, however, would not result from the type of system described here. Present information on the Wolof shows that in the past Wolof women could own slaves and cattle, but there is not enough information to show whether women also cultivated their own field. In Thysse-Kaymor and Sonkorong (regions of Sine-Saloum), it appears that at the end of the nineteenth century women owned the cloth woven from the cotton cultivated as a group on the communal cotton plot. Although most cloth was used to clothe members of the household, they could trade the remaining material for small cattle or ornaments. According to my informants, the material was the women's property and they could decide how much they would give to their husbands.[4] Around the turn of the century, the commercial houses started to buy groundnuts and to sell European-made cloth in the region. The lower price of imported cloth compared to that of local cloth led to abandoning the cultivation of cotton in favor of groundnut cultivation. The increase in groundnut cultivation also meant that groundnuts replaced millet and sorghum on communal household plots, as well as on plots belonging to young dependent males. Since cotton cultivation was abandoned, women also started to cultivate groundnuts in order to buy imported textiles and other articles. Small plots were cleared for them or vegetable gardens were enlarged.

So we see that Wolof women did not remain idle when cotton cultivation was abandoned. They became engaged in cashcrop farming which they still do. Besides planting their own fields with groundnuts, they earn additional income by performing agricultural activities on the plots of the men. By gathering and winnowing groundnuts they receive a present in kind ('wathan tal'), 5–6 kilogram groundnuts per day, equal in value to the local daily wage. Many African women are active in the trade of agricultural products: either foodcrops, in processed or unprocessed form from their own farms, or part of the cashcrop of their husbands. Putting these income opportunities aside, cashcropping by women themselves is more usual than generally acknowledged and is not peculiar to the Wolof. In the Sudano-Sahelian region it also has been reported for the Mandinka (Dey 1981), Serer (ICRA 1982), Bambara (Paques 1954), Sarakollee (Pollet and Winter 1968) and Haussa (Hill 1972; Belloncle 1980). It is also common in the Guinea area (Venema 1978).

In addition, Wolof men are still responsible for foodcrop production. The head of the household and male dependents clear the millet farm. They thin the millet clumps and weed twice in between the rows and on the rows. They cut the millet stalks and harvest the millet by cutting the ears. All millet is stored in a storage shed and it is the head of the household himself who hands out the daily quantity to be pounded. When the granary is empty, it is the man who is obliged to buy millet at the market. Women, in fact, only help in the sowing of millet. They also help in harvesting of millet by cutting the ear or in transporting it homewards, but they are not required to do so and if they do, they receive a payment in kind which is usually 5 kilograms of millet for a day's work.

A woman is, nevertheless, obliged to add, when it is her day of cooking, a quantity of millet equal to what is eaten by herself and her children. She obtains this quantity ('ndollah') by cultivating millet scattered over her groundnut plot, by participation in working parties at the harvest of millet, or through gifts from brothers, sons, and others. In addition to this obligation, she is also responsible for providing the herbs and relishes for the meals when it is her day of cooking. That men are responsible for providing most of the grain in food consumption has also been reported for the Haussa (Hill 1972, Raynaut 1977), Serer (ICRA 1982), Mandinka (Dey 1981), and Fulani (Dupire 1960). For cases in the Guinea zone see Venema (1978).

Projects of Agricultural Modernization

The greatest bottleneck in Thysse-Kaymor and Sonkorong to obtaining higher crop yields is the scarce and irregular rainfall. In the period 1969–1978, precipitation varied between 400 and 857 millimeters with a rainy season varying between 2 and 4 months. With a growing cycle of 3 months for millet and of 4 months for groundnuts, the precarious nature of agriculture in the area is clear. This problem is further aggravated due to the labor peak at the start of the agricultural season. In order to make full use of the wet season, it is necessary to sow as early as possible. However, millet and groundnuts are sown in the same period and it takes several weeks before all plots are completed if sowing is done manually. As a result, farmers obtain lower yields now than when they were able to sow immediately after the first rains.

Another problem is the degeneration of the soil. The field just behind the house is cultivated year after year with millet. Due to household refuse and the manure of the cattle which, in the dry season, is spread on this field, continuous cultivation is no problem. The groundnuts are cultivated farther away from the home and receive no manure. Therefore shortening the fallow period to only one year poses a problem.

Agricultural policy in Senegal promotes a permanent agricultural system with the use of oxen as the cornerstone. According to researchers and the extension service, use of oxen has several advantages. With the help of animal drawn implements, the sowing and weeding of crops can be done far more quickly than when done manually. The use of oxen, therefore, is a successful means of overcoming the labor shortage at the start of the wet season. Due to the stabling of oxen at the farm, manure is also obtained and can be applied to the groundnut fields. This manure and the use of the plow along with the introduction of oxen can improve the soil and its water retention capacity. The introduction of oxen in Thysse-Kaymore and Sonkorong met with great success.[5]

The general view is that agricultural mechanization results in loss of work for the women and consequently implies a loss of her influence. As is argued by Boserup (1970), and Goody and Buckley (1973), this is so because almost everywhere the use of the animal-drawn plow is the male's responsibility. Consequently, women are pushed back to the domestic domain and lose their position as producers. They also argue that modern farm tools are employed on the men's fields while the women are left with the traditional tools. But, as is demonstrated above, Wolof women grow their own plots of groundnuts and are not pushed back to the domestic work. Neither are they left with the traditional tools. As regards the latter, it had become the custom of the male dependents to help the women on their fields on the days when they were required to work on the communal farm. Because the work of the communal farm had priority, it often resulted in the women themselves sowing the groundnuts, only receiving some help at the weeding. With the introduction of the seed-drill and the cultivator, work on the communal fields is finished earlier. Instead of women sowing their groundnuts manually, this is now done by the male dependents in the household with the seed-drill. Also the weeding between the rows is done mechanically—the women only have to weed on the rows. However, the dates of mechanical intervention on the respective farms of the household follow the social hierarchy: first the farms of the head of the household are sown, then those of the male dependents, and at last those of the women. Because timely sowing and weeding is of great importance, the yields on the farms of the women are lower (Venema 1978:111).

Following the introduction of draft animals, it became the wife's duty to draw water for the oxen. Furthermore, she could be asked to lead the oxen. These two types of aid, however, are only requested when the head of the household cannot count on his sons. This happens especially in young households without adolescents.

Wolof women now work less on the communal millet field. While in hoe culture women had to assist in the sowing of millet, now this is done by men with the seed-drill. Up until a decade ago the introduction of the seed-drill increased the area cultivated by the women. Due to population pressure this is no longer possible, but a few women had been able to increase the size of their plots in such a way that they in fact make use of wage labor.

Mechanization here did not decrease women's income because it did not involve loss of activities women traditionally performed in exchange for a payment in kind. The introduction of the grain-mill in 1978, however, led to a loss of revenue because the women no longer could participate in working parties for pounding millet. The mechanization of crop processing activities via nut-cracking machines, oil presses, and rice-threshers has also resulted in a loss of income for women in many areas.[6]

State Marketing and Village Cooperatives

It often has been rightly observed that the efforts of the colonial and of the recently independent governments to increase agricultural production were restricted to the export sector. The village cooperatives in Senegal, established since 1960, only gave production loans according to the amount of groundnuts sold by the farmers. This has had a negative effect on the cultivation of foodcrops because the men have increased their acreage cultivated with groundnuts at the expense of millet production. In Thysse-Kaymor and Sonkorong in 1969, only 0.18 ha. foodcrops per head was cultivated, amounting to about 140 kg per head which resulted in a shortage of grain in the wet season. It was only from 1975 onwards, when the cooperatives started to buy grain, that the area cultivated with foodcrops increased.[7]

Boserup (1980) opposes the production of foodcrops by men because it could result in the loss of the little economic independence the women had. However, because this view apparently is based on those societies where women till only foodcrops, it is not correct for the Wolof.

With regard to the general consequences of official trade in foodcrops for consumption at the village level, it appears that the men increased production not only to sell to the cooperative, but also retained more than the traditional amount in order to speculate later in the season when millet prices increased. Although I cannot prove it, it is reasonable to assume that the greater amount of foodcrops at the village level has improved food consumption. Another advantage of increased food production is that the number of working parties for harvesting has also increased and thereby it has become easier to obtain the 'ndollah' required by the husbands. A more general argument is that the increase in foodcrops has improved the rotation of crops and so indirectly has improved agricultural production.

Men are virtually the only members of the cooperatives. In 1979 in Thysse-Kaymor and Sonkorong, of the 249 members, only 4 were women. Consequently, women (and male dependents) can only obtain fertilizer and farm implements via the heads of household. It has been repeatedly argued that the women's limited access to local organizations is a bias in favor of male agriculture. However, here it did not exclude the tilling of women's fields with modern tools. Also, it is very questionable if membership in the groundnut cooperative is really an advantage for gaining access to production loans. My research showed that the orders made by the farmers to obtain fertilizer and implements were handled very carelessly by government officials, resulting in incomplete and delayed deliveries. Also, there is an under-utilization of farm implements. This is due to the fact that farmers have not grown fodder for the oxen, causing them to be too weak to draw the implements without long resting hours at the start of the wet season. Under-utilization of the oxen also results from the fragmentation of the Wolof farm into individual holdings. For these and other reasons (Venema 1981) the use of farm implements often has not resulted in an increase in production allowing the use of production loans.

More serious is the fact that women are dependent on the men to obtain their seed since groundnut seed is delivered through the cooperative. Often the women obtain their groundnut seed at an interest rate higher than that charged at the cooperative. Of the eight cases investigated, I found four women paying the official rate (25 percent), three paying 100 percent, while one woman reported that she had obtained the seed free of charge. Another disadvantage is that groundnuts delivered at the cooperative are registered under the name of the member (the head of the household), whether or not the groundnuts were grown by him. It is the man who receives the delivery slips and, when it is payday, it is the man who cashes the slips. Because all the women were illiterate, and only some of their children had attended school, the rumors I heard of heads of household giving their wives only part of the money owed to them very probably are not groundless.

Increase in Household Expenses

Boserup and others argue that African women not only do the many domestic chores, but also have to support themselves and their children. Viewing this heavy burden, she speaks of the "African type of woman." However, as already stated, women are not responsible for producing the main part of grain for food preparation. What are her responsibilities for other household expenses and what are those of her husband? A Wolof man is responsible for providing his wife with decent housing and a bed, a hoe and sickle, and new articles of clothing on the Islamic feast "Tabaski" (Aid el kabir), the payment of the taxes, and the main part of the bride-wealth of his sons. In addition, he buys the rope and the bucket for drawing water and salt. A woman is responsible for providing the herbs and relishes of the meals when it is her turn to cook, firewood, and the payment of about one-sixth of the bridewealth of her sons. She also buys the pulley for drawing water and the medicine used by her and her daughters in case of illness.

From this list it appears that a substantial part of the household expenses are not borne by the women. Nor are new charges always on the account of the women. For example, grain threshing is paid by the men (5 CFA francs in 1979) while grinding mills are paid for by the women (6 CFA francs). Also, the cost of modern medicine is shared between husband and wife, the women looking after themselves and the girls, the men looking after the boys. However, medicines used by the very young children are always bought by the women.

There is no doubt that in other domains the household expenses for women are on the increase. Due to population pressure, the area cultivated has increased and many useful trees and plants have disappeared or only grow far away from the village. Many women now buy at the village shops herbs and relishes they formerly gathered in the forest. These include pepper, *netetou* (fruits of *nete: Parcia biglobosa*), and "lalo" (juice of *mbep: Sterculia setiguiera*). Women also buy ingredients not formerly sold in the shops such as dried fish, tomato paste in tins, onions, and sugar. In polygamous families at the time of the survey, the woman whose turn it was to cook would spend 50–100 CFA francs in the shops above what she added herself. Besides ingredients, women now buy soap and matches while, in former times, soap was homemade and the fire was left burning continuously. The Wolof woman also takes charge of purchasing household articles. Today household articles are much more numerous and include not only handicrafts but also imported articles such as plates, glasses, etc. Even if these articles are bought with the husband's dowry, in case of replacement it is the wife who pays.

The participation of men in paying household expenses is not only typical for the Wolof. Many Hausa men give their wives lamp oil, ointments, powder, perfume, and henna on religious or family feasts, or even on each Friday. They also pay for the larger cooking utensils (Smith 1955; Raynaut 1977). A sharing of household expenses has also been reported for the Mandinka (Dey 1981), Bambara (Caughman 1981), and Fulani (Dupire 1960).

Implications for Agricultural Development Projects

The gist of our argument is the need for more knowledge of the local farming system before projects of agricultural development are assigned and implemented. Given the importance of a systems approach, understanding the different positions and roles in the household is crucial in project planning. Although it might seem that heads of households or women can be approached separately, in reality the other members of the household will also be involved in the proposed innovations. In our view, if income increase is obtained through contact by the extension service with the head of the household, the additional income will be disseminated to a large extent to other members of the household. Nevertheless, a slow but steady structural improvement in the position of the head of the household [relative to the rest of the family] may result if the extension service limits its contact to him.

The position of women compared to the men is unequal in many ways. It is therefore a pity to observe that so little attention has been paid to women in agricultural projects. Dey (1981) shows that in three rice development projects in Gambia double cropping rice was introduced only for men, while rice was the traditional crop grown by the women and

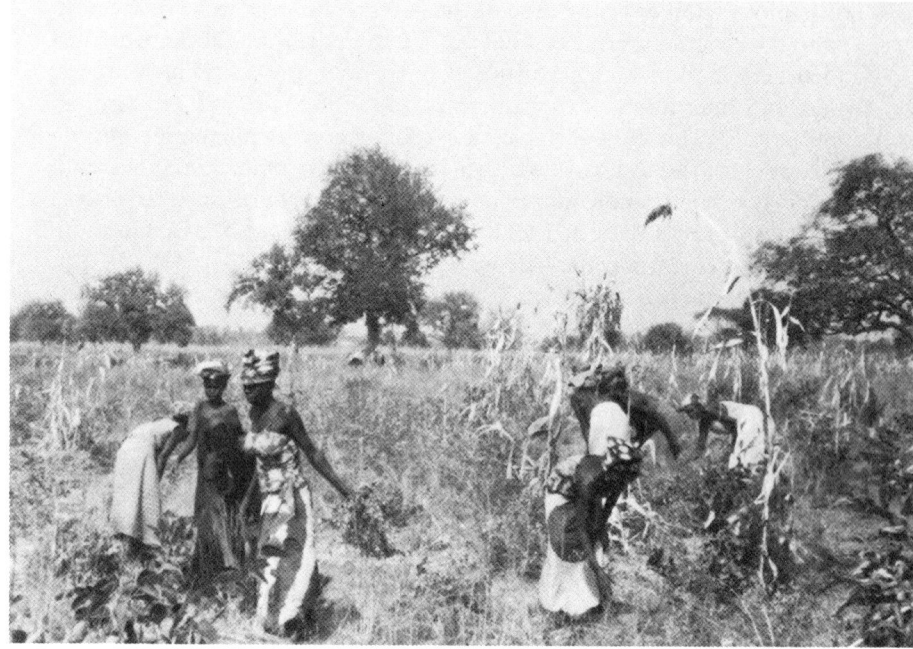

In this picture, Wolof women in Senegal are harvesting grain. "A rural woman participates in virtually all the work in the fields. It is not even rare for her to undertake this work alone if the men are absent" (Cisse 1983). Photo by Bernhard Venema

during the wet season women had more time to occupy themselves with rice because the men were already busy with growing groundnuts.

A main goal of agricultural projects for women should be to improve their material position which will, in our view, increase their influence within the household, allowing them to improve their lot and that of their children. For the societies reviewed here, to improve the financial position of women by teaching them food production is contrary to their own interest and a break from activities which always have been their task. Contrary to what generally is accepted, more attention has to be given to training women in cashcrop farming. One of the experiences of the *Animation Rurale* project in the region of Zinder, Niger, is that women were very interested in improving their groundnut cultivation. Therefore the extension service undertook to teach women such topics as improved seed, fungicide, planting distance, and so on (Belloncle 1980: 91). Therefore it seems very important that the *animatrices* dedicate sufficient time to teach women about cashcrop production. This may involve a change in the curriculum at the centres where the *animatrices* are trained.

In agricultural projects, more attention also has to be paid to which person in the household will be charged for the additional expenses resulting from the project. As I showed elsewhere (Venema 1982), the fact that men are paying for grain threshing seems not to be the result of a well considered agricultural policy but a coincidence. For the Wolof women it seems worthwhile to promote the growing of vegetables and spices [which have considerable market value as well as adding to the family food supply]. Along with this, extension agents have to explain to the men the medical reasons for supplying the household with sufficient grain; this perhaps also implies additional training for the extension agents.

Notes

1. 410 CFA francs = $1.00.
2. The cultivation of a personal plot under the obligation to work five or six mornings per week on the communal fields seems to have come into existence parallel to the rights of the domestic slaves to work for themselves after having worked for his master (See Meillassoux 1971, 63–65; Pollet and Winter 1968). The right of the members of the household to till a personal plot in exchange for labor obligations on the communal farm is usual in many (traditional slaveholding) societies such as the Serer, Toucouleur, Bambara, Sarakollee, Mandinka and Haussa.
3. Until World War II, male dependents worked five mornings per week on the communal farm in exchange for the usufruct of a personal plot. Since then they work only four mornings. The reason for this is that the seasonal sharecroppers ('navetanes') from Mali and Guinea only agreed to work for four mornings on the communal fields of the farmers in exchange for a personal plot. The farmers gave in, whereupon the male dependents followed the *navetanes*' example (at least in Saloum). For the precise contents of the rights and duties between the head of the household and male dependents see Venema 1978: 107–123.
4. The right of a Wolof woman to the locally made cloth was based on her activities in the cultivation and processing of cotton: she sowed and picked the cotton and spun and dyed (the weaving being done by the slaves in the village). Two generations ago, Bambara women in Markala spun and dyed cotton but the cloth woven was distributed by the men and only

the remainder could be sold by the women. This male ownership of cotton probably was related to the fact that here the cotton was provided by the men (Caughman 1981).

5. The number of farm implements and draft oxen or cows in the villages Thysse-Kaymor and Sonkorong (183 households in 1973)

	1969	1977
seed-drills	150	345
cultivators	117	263
poly-machines*	8	94
pairs of oxen/cows	25	175

Source: Benoit-Cattin 1977.
*A poly-machine is a multi-purpose implement able to sow, weed, hill up, and harvest groundnuts and to plow.

6. *Editor's Note.* If the women do not own the mill then they get no revenue from it, although it may mean they do not have to work as hard.

7. Area cultivated in food crops compared to the total area cultivated in Thysse-Kaymore and Sonkorong.

	1975	1978
% groundnut	61	48
% cotton	7	4.5
% foodcrops	32	47.5

(millet, sorghum, maize)

Source: Niang and Richard 1978.

SECTION II

Development Programs for Rural Women in Mali and the Sahel

Commentary

For at least ten years the governments of the Sahelian countries have voiced concern about women in rural areas. Sometimes this concern is focused upon the plight of women whose days consist of unrelieved and unrelenting toil (see Cisse 1983). At other times, interest is specifically directed toward increasing food production by varying what is planted and improving agricultural techniques thereby affecting women directly or indirectly. On other occasions, government officials are concerned with the general health and well-being (including literacy and basic skills) of the rural population, the majority of which is women. Numerous government programs, with and without the help of foreign assistance, have been established to address these issues.

The major obstacle to success in all such programs is lack of sufficient resources combined with the difficulty of altering the traditional social system. For example, if there were enough money to provide every village in the dry zones with a deep bore well so that water was not a problem, village women would have less work and all villagers would be more productive. The increased wealth would allow a margin for varied activities and facilitate the introduction not only of vegetable crops but also of grain crops which depend on irrigation.

In reality, this is not possible. A machine-produced deep bore well costs thousands of dollars, money which is certainly not available for all of the thousands of villages in the Sahelian countries. Even improving traditional wells—deepening, cleaning and making platforms for them—is expensive and not reliable as such wells often dry out.

In addition, it is not only the absolute lack of resources but the way people have traditionally behaved and their reluctance to doing things differently (because of lack of experience of benefits from changing their patterns of activity) which makes it so difficult to improve the lives and productivity of rural women. This reluctance cannot be eliminated overnight. Nor do the Sahelian governments want to destroy the traditional social system. Their strategy is to assist rural people to gradually understand different new programs: why literacy is important even for women, why children must be vaccinated, why food must be properly cleaned, and why selected seeds must be used and properly fertilized.

There is no easy way to do this, although it is very easy to find things to criticize in what has been done. Mistakes have been made and progress is slow. It is true that national politics in all Sahelian countries is controlled by men and the traditionally male-dominated social order is still reflected in the distribution of power in society. There often is one minister who is a woman, but she is most likely in charge of some social domain related to women. This power relationship is translated to every level of politics. In the village it is still the Headman and the male family head who make decisions and speak for their groups. Given this, and the overall poverty of the countries, it is not surprising that equalizing the position of women in society—and in rural areas in particular—has not received top priority. But, even if it had, it is not clear that the governments would have been more successful in their rural development programs which affect women.

In Mali, two major groups have focused primarily on working with rural women. The first are the ministry officials in charge of promoting women especially in the Ministries of Rural Development, Education, Health and Social Affairs, and on the Cooperatives Board. Many of these positions to promote women were created in the government in 1979 (Dicko 1983). The second and older group which pinpoints attention by the government and private citizens alike on all matters concerning women is the National Union of Malian Women (UNFM) which has been active since the 1960s. With money from the government and grants from donor agencies, UNFM runs a wide variety of programs for rural women. Although at least one observer (CILCA 1983a, 41), has criticized the lack of coordination and cooperation in women's programs, the leading women in UNFM and the government appear to be well aware of what each is doing for and with rural women (Interviews 1983).

There are six major activities supported by government and UNFM programs, sometimes combined in some way within a single project. The activities include: (1) setting up income-generating activities such as weaving, dying, cloth producing, or pottery making, and marketing the products; (2) establishing income-generating activities which also increase food production such as cultivation and marketing of truck garden crops; (3) alleviation of the overall domestic burdens of rural women through improved labor-saving technologies—smokeless wood-conserving stoves, new or deepened wells, and simple grain grinding mills; (4) general "education for living" programs in nutrition, health, and child care education (combined with other health campaigns); (5) adult literacy education; and (6) mobilization/awareness campaigns to enable rural women to undertake their own development programs, including setting their own priorities and designing and implementing goal-oriented projects once goals are clarified.

Illustration of the Range of Programs for Women

The seven papers included here illustrate the range of programs and activities in Mali and the Sahel. Mme. Traore's paper, for example, discusses the UNFM training center at Ouélessébougou which provides training for extension agents working with rural women in the second region of Mali, 78 kilometers south of Bamako. She discusses the preliminary stages involved in setting up the training programs and goes on to outline the methodology used in the development of training courses and the evaluation of participants. She shows the importance to UNFM of gaining local support from regional and village leaders and the complexity of UNFM programs as they train in health subjects, nutrition, vegetable and grain cultivation, and basic literacy.

Both Mme. Kantara and Dr. Henderson (in the papers which follow the Ouélessébougou presentations) discuss women's *animation* programs which are part of larger livestock projects in Mali and Upper Volta. The two livestock projects were funded by different agencies (Saudi Arabian Development Aid, FAO, and some USAID funds for the women's project in Mali, and USAID funds in Upper Volta)[1] and are described from a slightly different perspective. Mme. Kantara talks about the aim of her group to "find with our women farmers the solution to the squandering of energy (through hard labour, water duty, arduous field work) . . . ways of easing their burden and finally . . . a more satisfying meaning to life (by improving income . . . health [and] improving the way of life in the village)" (p. 120).

More cautiously and critically, Dr. Henderson describes her project as one "to encourage women's groups . . . to become involved in the process of development, . . ." adding that "at the time of the project, which was in 1978–1979, the ideas of grassroots committees and 'development from below'" were popular but that "the implications of such ideas . . . had not been carefully examined" (p. 134).

Both authors describe a process of initial research on economic, ecological, social, and cultural factors in the project areas. After gathering preliminary information, staff on both projects met with women in the villages to describe what they were doing to rouse enthusiasm among the local women and to begin the process of getting them to define their problems and their own solutions. Mme. Kantara's project defined a very broad role for the *animation* team while Dr. Henderson worked on a more limited mobilization effort. Dr. Henderson's project brought women's groups together and helped them discuss their problems and what they might need to solve them. Mme. Kantara's project team was not limited to discussing with women how to better handle livestock, agriculture and/or other income generating activities. Her team selected pilot villages where women were chosen by the villagers to be trained as *animatrices* and in which a contact committee was established with four members, each in charge of a specific area: (1) information and communication, (2) health and nutrition, (3) family economy, and (4) general labor reduction. The project introduced improved stoves and motor driven grinding mills to villages after training agents to teach people to use them. It made available six carts and six donkeys on a loan basis, helped set up vegetable gardens, and trained a poultry agent for a proposed poultry scheme. Moreover, Mme. Kantara's women's project will benefit in the future from the larger livestock project through programs to vaccinate the women's animals and teach them about animal care, milk processing, and conservation.

Mme. Djire's brief paper describes the work of the women's functional literacy program in Mali, begun in 1976, which uses literacy training to promote broad improvement programs in the areas of family care, agriculture, animal husbandry, handicrafts, trade, and income-generating projects. Five hundred women in the Koulikoro region alone have been involved in the literacy program and 30 women's centers in the Koulikoro, Segou, and Mopti regions of Mali have been established. Functional literacy programs train adult women to read, write, and cipher at a minimal level in their native tongue. What they are taught to recognize and use is directly related to the tasks and obligations of their daily life. Mme. Djire concludes: "Literacy programs teach women to read and write through themes pertaining to health, agriculture, handicrafts, education etc. . . . and teach them to better understand the meaning of modern life, to adapt to their changing environment and increase their productivity" (pp. 155–56).

The task of the women's division of the cooperative board, described by Mme. Diallo, is more specifically directed to a single objective than the previous programs described. The division seeks to facilitate the development of cooperatives in women's productive activities. It mobilizes women to participate in cooperatives in agriculture, animal husbandry, wood collection, fishing, plant gathering, handicrafts, and savings. Cooperatives in gardening, dyeing, and pottery already have been set up in rural communities while sewing, soap, and agro-pastoral cooperatives have been set up in the Bamako region. The program is an attempt to educate women in cooperative principles and help them with common problems such as finance, management and administration, marketing, and supplies.

The final area of activity illustrated by the papers is the introduction of improved stoves, reducing the amount of wood used and time needed for gathering fuel and cooking. Mme. Ki-Zerbo and Mr. Tucker are talking about the same Sahelian program, with specific reference to Upper Volta, which used a large, relatively difficult-to-construct cook stove: The progenitor of this stove was developed in India and adapted for use in other areas of the world. It not only can reduce labor time but also can help with the increasingly severe deforestation problem in the Sahel and may facilitate better cooking and reduce certain hazards such as children getting burned.

Mme. Ki-Zerbo discusses various aspects of her work with this program including teaching men and women *animateurs* and *animatrices* to build such stoves. She emphasizes the importance of the stove program and of introducing other labor-saving technologies based on existing technologies and developed to suit the self-perceived needs of rural women. Mr. Tucker provides a more critical perspective, talking about major problems with the stove program and suggesting that the achievements of the stove programs in Upper Volta and the Sahel have been limited. The two papers illustrate the hopes engendered by such programs in "appropriate technology" and some of the drawbacks which they have encountered in the Sahel.

Conclusion
The Record of Projects for Rural Women in the Sahel

In most cases, since the programs discussed were ongoing when the papers were written, a final evaluation is not available. However, there is some evidence available on project performance from other Sahelian projects for women which is not altogether encouraging.[2] Thus, despite lip service given to the need to have rural women enthusiastic and fully informed about projects, many programs suffer because the women do not really understand what is happening. Undoubtedly they were enthusiastic about the new programs which would increase their options. But outsiders decide what will be done—outsiders who are members of an intellectual elite. They are far removed from working in the fields and make judgements using criteria the village women do not have. In one Malian project, for example, the government sought to introduce women into the training programs of Centers where selected farmers are given brief training programs. In the first round of the project, observers found that men already at such centers (Centres d' Animation Rurale [CAR]) did not understand the necessity to train women and that the women initially chosen did not really know the purpose of their participation and were not highly motivated (Van den Oever-Pereira 1979). Similar problems were encountered when women were trained for the first time at the training center for extension agents (the Centre d'Apprentissage Agricole [CAA] Schoepf 1979).[3]

The American Friends Service Committee organized a woman's project (in collaboration with UNFM) in the First, Second, Third, Fourth and Sixth Regions of Mali. The program supported activities (in seventeen villages) such as cloth dying, gardening, and soap making. The program had five major parts: (1) technical assistance to groups, (2) a training program for women, (3) marketing of craftwork through the shop, La Paysanne, in Bamako, (4) financial aid to small production units, and (5) research into appropriate technology for village production. This project tried to avoid some of the obstacles to other women's projects—there was considerable effort to keep the women participants closely involved in decision making and in taking responsibility for the various facets of the projects's operation. But the 1981 evaluation still reported certain problems—some groups were much more successful than others. "A certain discouragement is apparent among some early groups. When asked, 'Why are you a member of the group?' they reply unanimously, 'For the money.' In some cooperative groups, not only do the women receive nothing, but they do not even know what or how much the profit is. They have shown their discouragement by progressively abandoning the groups" (American Friends Service Committee 1981, see also Coulibaly 1981). Furthermore, although the project encouraged cooperatives, village women were not interested in cooperatives (except perhaps for marketing or obtaining inputs); they wanted their own plots of land.[4]

Poor understanding by village women is compounded by numerous other difficulties such as shortage of supplies, poor marketing and credit arrangements, faulty analysis of what economic or agricultural activities are really sustainable, inaccurate assessment of obstacles including pests, drought, lack of village leader support, and unequal treatment by responsible government agencies.

Such things are graphically illustrated in a report on women's projects financed by USAID in Senegal (see Jeffalyn Johnson 1980). The firm of Jeffalyn Johnson and Associates, in an analysis of a project in the Casamance (Senegal) to help set up women's cooperatives for vegetable production and marketing, concludes that "the Casamance project was, at best, a marginal success" (Appendix 1, page xi). Beneficiaries had not requested the project, inputs and treatment by government officials varied greatly among villages, and the relevant government agencies had not coordinated with each other (Jeffalyn Johnson 1980, p. x). After examining the second project in Kassack Nord (Senegal) involving a broad range of economic activities at the village level, their conclusion is slightly more positive. Yet efforts to improve vegetable production had failed and the other projected outputs had not (at the time of the evaluation) been realized (Jeffalyn Johnson 1980, Appendix 2, pp. x–xiv). The third Sahelian women's project in Tivaoune (Senegal) was more positively evaluated but also suffered numerous problems including lack of necessary *animation* and training for the women and inadequate assessment of the proper kind of tree to be planted for firewood and the most appropriate form of sheep pen to be used by the village women. In regard to food crop production, the growth of manioc and *niebe* (basic food crops) had been proposed but manioc was not planted because of fear of insect destruction, and a cash crop (peanuts) was substituted. Niebe, however, was produced as projected. In this project, too, outputs projected were not achieved as scheduled (Jeffalyn Johnson 1980, Appendix 3, pp. vi–viii).

This type of problem and this kind of success record are common for Sahelian women's projects in general. These projects and their track records provide a learning experience. It is true that project failure has a high human cost, but doing nothing has a higher price. Serious efforts to improve the lot of rural women are quite recent. The whole process of rural development is so poorly understood that it would be unreasonable to expect that mistakes would not be common.

Probably the most impressive feature of the papers that are included here is not the unresolved issues or the mistakes, but the overwhelming enthusiasm and commitment of the African women. With little experience themselves, they are pushing ahead to try to turn what went wrong to a better direction and to build real improvements with, and for, rural Sahelian women.

Notes

1. FAO = United Nations Food and Agriculture Organization
    USAID = United States Agency for International Development
    The larger stock farming projects were funded by a combination of donors. The

women's projects were particularly interesting to USAID which funded them in both countries. The Malian stock farming project still exists. The Upper Voltan project was not officially abandoned, but stagnated after USAID did not renew its funding.

2. Perspectives on this performance record may change when studies by Dr. Kathleen Cloud and USAID of past women's projects in the Sahel are published (expected 1985). These studies should provide important information for scholars and planners to assess what has been done and how different strategies have worked.

3. Evidence suggests that the initial problems due to lack of understanding by women participants (or their male family members) of the importance of training women at these centers has diminished. Both centers currently are training women who are thereafter employed as extension agents and *animatrices* by the government. The number of women who have actually completed the program is not available.

4. Telephone conversations with Patricia Hunt of the American Friends Service Committee, October 20, 1982.

See discussion on cooperatives in the Editor's Note to Mme. Diallo's chapter.

5 The Ouélessébougou Training Center for Rural Women Extension Agents

Halimatou Traore

Editor's Note

Mme. Halimatou Traore is director of the Training Center for Rural Women Extension Agents at Ouélessébougou. This center is run by the Union Nationale des Femmes du Mali (UNFM), the national woman's organization which is the chief advocate for projects and programs to improve the lives of women in the country.

Ouélessébougou is in the Kati *cercle*, fairly close to the capital, Bamako. Mme. Traore's discussion of the center, opened in 1980–1981, is an interesting description of a training program for village extension agents (called *animatrices*). Several points make this chapter particularly noteworthy. Mme. Traore makes a very clear presentation of how the Center first began its work, how it set up its training sessions, and how these are run. She also shows the efforts of the Center's staff to observe (and respect) local customs and preferences while training women in production as well as domestic and health-related activities. The emphasis in her presentation on *animation* in the sense of "awakening" and mobilizing local women as the key to success in rural development is important. Mme. Traore's own commitment to work with rural women is revealed in the paper and is a testimony to the dedication of many of the other women presently running projects for rural women in the Sahel.

Introduction

The Union Nationale des Femmes du Mali (UNFM) is the only women's organization open to any Malian woman regardless of her race or religion. UNFM's mission is to establish ties of friendship among women who belong to all social strata of the country, to defend the interests of women and the family, and to enable women to participate more fully and more efficiently in the economic, social, and cultural development of the country. Among UNFM's tasks are the following: (1) to raise the civic and political consciousness of Malian women so that they become well-informed guardians of rules and regulations guaranteeing the stability of the nation, the cohesiveness of households, and Mali's cultural and moral values; (2) to fight for the emancipation and promotion of Malian women and prepare them to play a dynamic role in the development of the country; and (3) to lighten the domestic tasks of Malian women and develop their capabilities in order to steadily increase their role in the production and management of community affairs.

To carry out its mandate UNFM has undertaken, among many other projects, a program to train women extension workers. This project originated after a visit to the Republic of Mali in 1975 by two women belonging to the "Black Women's" organization in the United States.[1] On August 1, 1980, a grant agreement was signed between the Government of the Republic of Mali and the United States of America to start the project. Thereafter, the Government of Mali placed the project under the direction of UNFM. An interministerial coordination commission composed of representatives of the Ministries of Health, National and International Cooperation, Planning, and Rural Development, and of the Social Affairs National Board, the Rural Animation Training Board, the Functional Literacy Board, and the Koulikoro Governorship, was set up to ensure the proper functioning of the project. The Training Center for rural women extension workers, known by the initials CFAR-UNFM, is located at Ouélessébougou, in the second Region of Mali, 78 kilometers south of Bamako.

The objective of the CFAR-UNFM is to develop the body of knowledge about rural women, lighten their work load, decrease the illiteracy rate, and give women a greater opportunity to play an active role in the rural economy. Our Center offers to rural women courses which can lead to an improvement of their productivity, health, and literacy level. In short, its aim is to help rural women improve the quality of their lives at the village level by increasing their knowledge in matters that are of direct concern to them while taking into account their beliefs, traditions, and know-how. The Center works with the traditional authorities and all existing structures and takes advantage of its own potential to facilitate the transmission of the message.

Project Area

Ouélessébougou is the chief town of the Kati *cercle* with a population of 33,000 and an area of 2,550 square kilometers.

It was selected as the location for the Training Center for rural women extension workers, largely due to its closeness to the capital, its ease of access, the open spirit of the population of these villages, their frankness of expression, and their easy adaptation to changes. These are my personal opinions . . . there may very well have been other reasons. I also think that the action of our good Belgian Missionary Sisters in Ouélessébougou over the past ten years is largely responsible for the positive open outlook of the population.

The village of Ouélessébougou was founded about 800 years ago by a man named Ouelesse. His eldest son, Fassoun, was the first village Head. Nine years later, Ouélessébougou N'Tintou, where the center is now located, was founded.

At the beginning, the population of Ouélessébougou was made up of three large families: the Samake, the Bayayogo, and the Coulibaly. Later on other families joined them including the Diawara, the Goumbia, and the Soumaoro. The district of Ouélessébougou is made up of seventy-two villages. It has thirteen sectors: Beneco, Ferekoroba, Tinkele, Ouélessébougou, MPiebougou, N'Tintou, Simidji, Marako, Dialakoro, Digan, Sanankoro Djitoumou, and Dierra. The main ethnic groups are the Bambara, Samogo, Mossi, and Peulh. Existing religions include Islam, Catholicism, and Protestantism although some residents remain fetishists.

The various economic activities of the district are agriculture, cattle breeding, trading, and processing (*soumbala, datou,* [Editor's note: those are traditional foods processed from wild plants] and traditional dyeing and spinning, etc.). The High Valley Operation [the major government program] covers the entire district of Ouélessébougou and works primarily in agriculture, principally in cereal cultivation. A family planning program has also been established to ensure food self-sufficiency [by reducing demands on food]. Large scale cattle breeding is practiced by

Peulhs and sedentary peoples. However, the scarcity of equipment . . . and the wandering of animals hinder veterinarian services. The Catholic Mission's help is required to preserve the vaccine and the government is called upon to control cattle wandering. From May to December [during the wet season], pasture land is sufficient, but, for the rest of the time, the cattle are taken to the Keleyadougou region to seek water holes.

Government Services

A forestry agent takes care of all problems concerning the forest, fauna, and water at the district level. There is a Rural Animation Center (CAR) belonging to the Government, established in 1968. Until 1974, it was only for unmarried men; thereafter it became a mixed Center. Couples and unmarried men and women receive training there for two years. In addition, a cooperative extension agent works with the people and prepares them to organize cooperatives.

Primary schools exist in all sectors of the district. In the Ouélessébougou township there are three schools, two of which offer both primary and secondary education; the other offers primary schooling only. There is a dispensary and childcare and maternity centers. In some of the larger sectors there are also rural maternity centers. The personnel of these centers is composed of Malians and Belgian Sisters. The Belgian Sisters also take charge of the malaria prevention program in certain villages. The district has a police brigade to maintain law and order at the local level, a branch of the Popular Pharmacy to supply medicine, and a branch of the Somiex to provide essential goods. The District Government Office is composed of the Head of the District, a Secretary, general clerks, and keepers. In the political field, all democratic organizations are represented including the Malian National Party (UDPM), UNFM, and the youth organization (UNJM). These various organizations play a very important role in the development of the district.

Customary Authorities and Practices

The current Head of the village, Mr. Zina Samake, is a direct descendant of Ouelesse. When we met the leading citizens, they also spoke to us of the traditional association, called "NTogon", which organized the village elders who traditionally were the ones to sit under a tree and discuss various social problems such as marriage, baptism, and divorce. Now this association exists under the name of "Dougoutiguiboulon." "Tons," traditional farmers and cattle breeders associations, also exist at the village level. In certain villages (Sounsounkoro, Marako) there are some associations exclusively for women farmers. There are also women's associations divided by age groups. Young girls, once married, leave the group.

Betrothal occurs at the age of three or four years. The girl is practically brought up in the boy's family; she only rejoins her own family during the dry season. However, during this period of coming and going, she does not know [sleep with] her betrothed. Excision [clitoral] occurs during the year of her wedding. In addition to traditional groupings, there now are other rural development groups at the village level. A Primary Federation organizes all the rural groups at the district level. There is also a Parents' Association (APE) which takes care of the construction of schools, maternity centers, and social centers.

At the outset, the project staff sought to get acquainted with the project area and its people. We began by making contact with representatives from the various social, administrative, and political strata of the district of Ouélessébougou. Then, meetings were organized with the traditional administrative and political authorities, and with representatives from local government services. These various meetings enabled us to gather data on the demographic, administrative, and political situation and the existing technical services operating in the zone.

After undertaking this study, much time was spent on mobilization of the local population. Once the round of visits to the seventy-two villages (composing the district of Ouélessébougou) was made, several "awareness" campaigns were undertaken in the twenty-five villages chosen as starting points for our action. During the course of these programs, we became better acquainted with our villages and we were able to list the needs of our sisters and then, together with them, analyze these needs to try to find the solutions which would satisfy them.

The Mobilization Program

The first tour of the seventy-two villages of the district of Ouélessébougou lasted twenty-five days. There were six people in the delegation: the Head of the Men and Women's Animation Center (CAR); two instructors (women) of the CFAR-UNFM; the agriculture instructor of the CFAR-UNFM; one member of the local Women's Union; and the chauffeur of the CFAR-UNFM.

During this first tour, all villages were informed about the establishment of the CFAR-UNFM at Ouélessébougou, its objectives, and the method of choosing trainees from the villages. They were also told about life at the Center, about the program in general, and about the probable length of the sessions (1–3 months). The schedule for training for the seventy-two villages (twenty-five villages to be trained per year) was also discussed. Finally the villagers were informed of the construction schedule and the probable opening date of the Center.

Wherever we visited, we asked to meet the entire village: the Village Head, his Councellors, the family heads, the representatives of youth, and all the women. This first tour enabled us to pass through all seventy-two villages of the district and to better understand the attitude of the population (which made a fairly positive impression at the start). We also saw the structures which already existed at the village level: the schools, the maternity centers, and other rural groups. We met the agents already working with the local people with whom we would have to work such as the Head of the Zone, the Head of the Base Sector, teachers, male nurses, matrons, literacy instructors, etc. Thus, this first tour gave us the opportunity to have a preliminary contact with the villages.

It should be noted that in spite of their enthusiasm for the creation of the CFAR, men sometimes expressed some worry concerning the time their wives would spend at the Center.

After this first tour, true, in-depth, mobilization campaigns were undertaken. The members of the delegation for this campaign consisted of two women instructors and the CFAR chauffeur. We waited until evening to arrive in the villages in order not to disturb the many daily tasks of our sisters. They then would come to us in great numbers. But rural women do not like to talk in the presence of men. That is why, after having asked the Head of the Village's permission, we would meet only with the village women. We then would have a lively conversation centered around the pictures of the education series, "villagers come to life", which we brought with us.[2] The use of pictures not only amused our sisters, it also has helped them to slowly express their problems. Together we analyzed the problems and sought ways and means to solve them. At first this type of work was fairly difficult. But now our village sisters are more or less used to analyzing problems and seeking solutions.

Training

Starting with the needs of our sisters at the village level, we defined the themes of our different training sessions. In order to do this, a commission composed of personnel from the various technical services (Nutrition, Health, Education, Cooperatives, Family Health Division, women's division of the National Functional Literacy Board, National Rural Animation Centers Board, Hygiene and Sanitation Service, High Valley Operation, etc.) was formed. Courses are given in Bambara (see methodology section). The length of training varies from fifteen days to a maximum of thirty days. The CFAR has a boarding program and trainees are fed and housed at the Center. Each trainee can come with her youngest child (aged from infancy to six years) as a day nursery exists at the center. Each village sends us two trainees per session.

The selection is made at the village level by the traditional authorities in cooperation with the local women's organization of the village. The following criteria have been used (although they are subject to revision): (1) Only married women capable of transmitting what they have learned to the other village women will be chosen. These married women, with or without children, should be respected and should have a certain influence in the village; (2) the number of children will not be taken into consideration; and (3) at the end of their training, the women receive the job of *animatrice* at the village level in the field of their training.

From November 1982 until June 1983, 171 trainees were trained in health, nutrition, vegetable production, infant and maternal care, and functional literacy.

Between June 1983 and September 1983, an agricultural program was held in the field for five pilot villages, the principal trainer of which was the CFAR agricultural instructor. The training undertaken is as follows: a third of the time spent on theory, two-thirds on practice (practical demonstrations, field trips, field work, etc.). During the training session, the trainees themselves clean their lodgings and prepare their food. They share in the work being undertaken (harvesting, picking cotton, threshing millet, watering gardens, watering trees).

Training is specialized. We prefer specialized extension workers to multipurpose extension workers because our sisters are illiterate and they therefore cannot take notes and can only rely on their memory. Multipurpose extension workers have too much to do and to remember at the same time—as the saying goes: "Grasp all, lose all." If one wants to teach them so many things at once, and ask them to put this into practice at the village level, it requires full time availability. But they have other responsibilities to fulfill and they are not paid for their extension services. Specialization lessens confusion and specialized lessons are easier to assimilate. There can also be several extension workers in a village, each one in her own field, which encourages healthy competition. Thus the training programs reach several women of the village rather than only one and *animation* becomes part of the whole village. The number of trainees per village is set at two per session so that they can discuss, help, and complement each other. Back in the village, they must work hand in hand. That is why we think it wiser and more useful at the beginning to have separate training sessions, each aimed at a specific theme.

Methodology Followed at the CFAR

Courses are given in Bambara in the form of lectures. They are made relevant through use of pictures, posters, charts, film projections, video, practice sessions, and field trips. The instructor takes care to group together all educational objectives belonging to the same theme and must prepare his lectures in advance. Before starting a lecture on a given theme, an evaluation of the knowledge of the trainees concerning the theme in question is made. Starting from this knowledge, questions arise from the trainees and then the lecture begins. After each lecture, a short evaluation by questionnaire is made to see if the message has been understood. Role playing is also used. It is an educational experience during which the trainees act out a real situation. This entertains the trainees and prepares them for their future tasks.

At the end of each week (or at the beginning of the following week) an overall evaluation is made, with a question on each subject treated. A final evaluation is also made at the end of the session by questionnaire.

At the end of each session, the trainees receive a personal allowance which varies according to the number of days spent—5,000 FM for fifteen days, 7,500 FM for twenty days, and 10,000 FM for the maximum period of one month.[3] They also receive a grant for the village. For example, after the session on health and nutrition, the trainees received some nivaquine pills, aspirin, a bottle of argyrol (eye medicine), Gomenol oil (essence of niaouli leaves, used for colds), bandages, cotton, etc. for their own use. Their village received a first aid box with supplies amounting to about

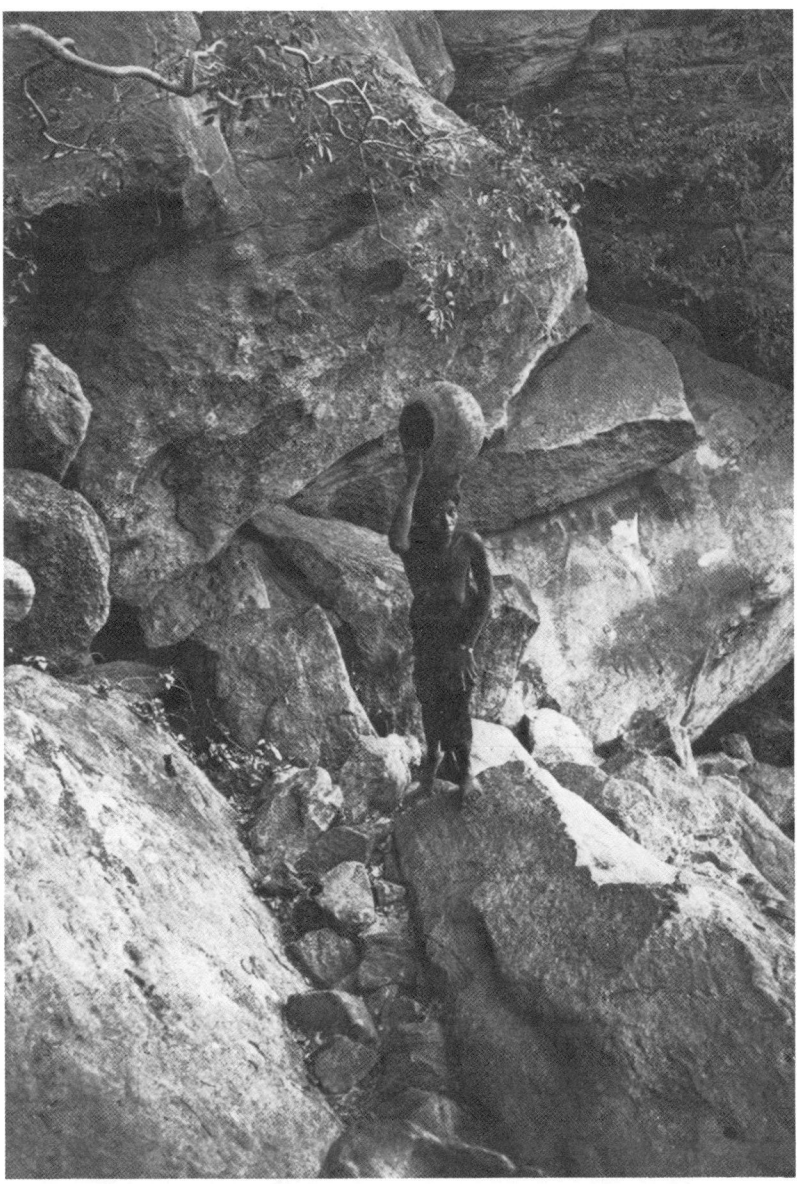

Hauling water up a cliff in Mali, this Dogon woman illustrates one of the most difficult and time-consuming tasks which women in the Sahel must perform at least once daily. Photo by Michel Renaudeau

26,000 FM. These medicines were to be sold to the villagers at reasonable prices. The amount collected is used to replenish the stock.

At first, the extension workers in charge of fund management for the village always pass through the Center so that we can see if the medicine is being properly sold and if the health program continues. Also, the trainees buy more medicine from us at a reasonable rate (they benefit from a small price cut). Now certain villages are already recognized by the Popular Pharmacy and their agents do not come to the Center any longer. They buy directly from the Pharmacy to renew their stock.

For the vegetable growing session, the trainees receive a personal allowance of seeds, insecticide, and chemical fertilizer. The village receives a cart to benefit the whole community. The money collected from its use is kept in the village and used for upkeep, repairs, and even the purchase of another cart or other equipment.

After training, traditional midwives receive a medicine box containing some medicine and small tools to be used for cutting, cleaning, and bandaging the umbilical cord (blades, alcohol, bandages, cotton, compresses etc.). Trained traditional midwives also receive 1,000 FM for each birth; 250 FM is for themselves and 750 FM for the village women's association fund.

As can be seen, our wish is that the village allowance be a source of income for the community and our main concern is the maintenance and the increase of this source of income. At the village level, the population has understood us—as the saying goes, "God helps those who help themselves." That is why our village women know that for all projects or mini-projects which could be of interest to them they must take the first step. They are concerned and want to participate in the preparation of projects. I am convinced that the best projects are those which come from the base—from those who will benefit directly from them as they are the only ones who really know their needs and their priorities.

Follow-up and evaluation occupies a very important place determining the impact of the training at the village level. It is useless to train people if they are not followed, sustained, and helped in the field especially. Thus the *animatrices* go to the village to see what they can achieve in their own milieu after their training. They encourage the villagers and congratulate them at every opportunity. The villagers who have been successful are used as an example. Villagers learn what others have been able to achieve and then they often decide to do as much or even more. In this way a healthy competition develops in the village. The follow-up enables us to undertake not only an evaluation but also a brush up of training of the village *animatrices*. During the follow-up, it is possible to ascertain in the field what the *animatrices* have been able to achieve with the people through practical demonstrations, lectures, etc., and the difficulties they have met in trying to put into effect what they have learned (the problems they have been able to solve and the measures envisaged to solve those which remain). Also, the pressing needs of the village can be assessed as can the steps taken by the village to satisfy them.

It is during the follow-up that we discover the most devoted villagers, those with the most initiative, who seek to solve their problems and who are always ready to help in the development of the village. At the moment, nothing is provided for these villagers and the Center can only offer them congratulations and encouragement. But we visit them regularly and if they have small problems at the village level . . . we help them to solve them. It is also for these villagers that we organize the agricultural sessions where they benefit from advice on the use of fertilizers and other materials. The same is true for small projects which we prepare with our villagers to help them meet certain of their needs. However, I honestly think that villagers who distinguish themselves by their devotion, their courage, and their spirit of initiative should get more attention.

At the end of the follow-up, the instructor writes a report. The different reports are examined and we plan our future actions in the light of the results obtained (possibility for retraining, insisting on certain points of the program, preparation of small projects to help satisfy certain needs of the villages). In September 1983, an evaluation of the project as a whole is projected with the full participation of the village women. Other women are accustomed to evaluation—each time we go to the villages during the follow-up we ask our women to tell us all that was achieved in the village since the day we came to the village.

Conclusion

While it is true that we are able to solve certain village problems by organizing training sessions at the Center, there still exist certain problems which we are unable to solve (water problem, grinding mill, etc.). It would be desirable to find a solution to help us satisfy these pressing needs of our villages which are beyond our scope and which can slow down our action.

Rural *animation* is not a profession like others. One must be constantly in the field, constantly in contact with the villages and one must like *animation*. It is truly a mission. It is only in this way that we will know the reality of our villages. In this way we can plan good, long lasting projects. Villagers themselves will determine their participation within the project. It is not possible to undertake development with projects prepared from the outside. Those who are to benefit from the project must be associated with the idea. It is even better if the idea comes from them and if they can question it. This creates a certain dynamic and gives the local population the opportunity to become creative.

By creating this Center, the Union Nationale des Femmes du Mali tried to accomplish one of its essential aims which is to improve the living conditions of our rural women. By so doing, it takes an active part in the socio-economic development of our whole country. The ideal would be that other Training Centers for rural women extension workers be created in all the regions of the country. The success of such centers will depend in large part on the efforts in *animation* which they carry out.

Notes

1. Editor's Note: This was the Black Women's Community Development Foundation.
2. Editor's Note: These visual aids depicted women at their various daily tasks and were used to provoke discussions about what is and can be done to solve daily problems.
3. U.S. $1.00 = 410 FM.

6 Training of Rural Women with the Stock-Farming Development Project in Western Sahel

Coulibaly Emilie Kantara

Editor's Note

Mme. Coulibaly Emilie Kantara spoke at the 1983 Bamako Workshop with great enthusiasm about the work of her project. She is employed by the Ministry of Rural Development to head the sub-project to promote the interest of women within a larger stock-farming project in the Western Sahel Region of Mali. Her chapter is very informative and provides an interesting contrast to the chapter by Dr. Helen Henderson which follows it. Mme. Kantara is uncritical and optimistic, seeing problems encountered as merely steps on the way. Dr. Henderson is more skeptical about methods chosen in her project and its overall impact although this difference in attitude may be partially due to the fact that all funding was discontinued in the Upper Volta project while the Mali program is ongoing.

Like Mme. Traore, Mme. Kantara describes the organization and method used by her project team. Her project sought to improve the life of rural women in production and domestic tasks and in terms of general well-being. She does not have the same emphasis that Mme. Traore had on teaching a single area to a village woman selected as *animatrice;* she talks about *animatrices* going to the villages to teach in all relevant areas. Like Mme. Traore, she places great emphasis on *animation* of local women and is very critical of past efforts which were too directive and did not really get rural women to fully participate. Perhaps the most noteworthy feature of Mme. Kantara's presentation, however, is her frank acknowledgment of problems encountered by the project including inadequate staff, the death of animals given on loan to village women, introduction of an inappropriate or expensive and complex domestic technology, and discontinued funding which stopped the work of the project in some villages.

Introduction

We know that we are at a turning point in development and training programs. Adult training methods now call for more participation and start with informative questionnaires. It is true that even well-financed projects initiated from the outside have never been assured success if they did not have the population's support. Now this is consciously recognized in the preparation of projects.

In the past, there has always been a certain lack of communication between the administration and rural people. It is therefore necessary to establish frank collaboration between the farmer and the new development agent and, in the long run, to be able to do without the development civil servant. To achieve this, long-term mobilization and training work must be undertaken to enable the farmer to take responsibility for his own development.

In light of this principle, the function of rural *animation* is to bring a group of farmers and development agents to think about their problems, analyze them, see the solutions already proposed, and find more efficient ways and means to improve development efforts.

Presentation of Our Project

Our project is called the "Stock-Farming Development Project in the Western Sahel" and is known by the acronym Pro-de-so. It is concerned with animal husbandry, one of the pillars of the Malian economy and one of its principal sources of income. Before the drought, the country had five million inhabitants, nearly five million cattle, and ten million sheep and goats. When primary production reached approximately 50 percent of the gross national product in the Malian economy, animal production represented 23 percent. After the drought, however, the situation deteriorated and animal production now represents only around 19 percent of the GNP.

Many coastal countries depend on Mali for animal production.[1] National efforts must be undertaken to examine changes in the standard of living, demographic transition, and changing consumer patterns in these client countries which modify eating habits and the demand for animal products.

The Saudi Arabian Development Fund and the government of Mali undertook the financing of investments in the two pastoral zones of Prode-so. The FAO participated in the development phase from 1980–1983. Briefly summarized, the project entails the development of the two zones. In the Nara-Est pastoral zone, the project will support the installation of equipment for well drilling, the reinstallation of village wells, the management and protection of the pathways for seasonal moving of animals, and the systematic improvement of animal health service. In addition, cattle breeders associations will be organized to help carry out the program and the management of the new installations. Village lands will be developed, and a temporary fund for local stock farming of cattle, sheep, and goats will be established. A technical *animation* and extension staff structure will be set up.

The project in the Kayes-Nord pastoral zone, whose first phase spanned five years (1978–1983), will consist of the progressive development of pastoral zones (including the creation of more water source points, both pastoral and village, pasture management, and the improvement of health services.) Breeders associations will be established to help carry out the program and to manage the new developments. Efforts will be made to improve the health of the population. A credit system will be set up for breeders and professionals in cattle fattening operations, for butchers to supply the Kayes market and for exporters. A three-level slaughter house will be built, and a sheep ranch, a cattle feed lot, and a breeding ranch close to the feed lot will all be established.

The FAO will provide zootechnical assistance and rural *animation*— in particular, identification and establishment of breeders associations in both zones. It will also assist with commercial technical management because of the particular location of the Kayes-Nord zone at the border between Senegal and Mali. In addition, the FAO will support *animation* for women in the zones. This technical support from the FAO during the starting stage complements the technical assistance provided by the Saudi Arabian Development Fund.

Rural Animation for Women: The Case of Pro-de-so

Our women's *animation* group first entered into contact with the project and with other projects sponsored by OMBEVI [the Malian government] to explain the necessity of creating a women's animation section. The term *"animation"* often made every one smile for, as used by the development expert, *animation* may have a very pejorative overtone. The term is used loosely. There are many ways in which *animation* has been used: directive animation first known in Africa through cash crop companies; religious animation; political animation; recreational animation; and youth animation (scouts, CAY, Catholic youth organizations, guides, pioneers).

We have a specific meaning for the term *"animation"* and have explained to our working associates the specific task that we wanted to undertake in our *animation* program. We intend to act by mobilizing the population to participate in all the decision centers of the project and, at the level of the local village, bring about changes leading to the improvement of life of our villages. We want to find *with* our women farmers the solution to the problems of the squandering of energy through hard domestic labor, hauling water, and arduous field work. We wish to find with them ways of easing their burden and finally, if they agree, to find with them a more satisfying life by improving income, improving health, and increasing the joy of life of the village.

There was no opportunity for us at the outset to address ourselves directly to village women. Before being able to contact them and start a dialogue, we had to go through the official and traditional channels that separate us from them. We therefore had to solicit and obtain the agreement of men. We had to see the regional authorities of Kayes, Koulikoro, and Segou, the authorities of the *cercles* within these zones, and the ward authorities. The traditional authorities of the village were then interviewed.

The whole long process of mobilization and awakening of rural women begins with men and especially our own husbands who, in Africa, are not happy to see a wife-*animatrice* always on the go. But we go on with the work, making courtesy calls, introducing our contact committees, and holding inter-service meetings to break the ice among the various partners. It is at the level of the regional authorities that it is necessary to get the strongest support. Governors and Heads of zones, if they are favorable to women's *animation*, will facilitate the actual implementation of field activities. We would even be able to call upon their help at the level of regional and *cercle* development committees. Project technicians and other (regional) development agents belonging to national services would help us with various administrative, technical, and social problems.

Administrative decentralization at the level of *cercle* and ward authorities can help us solve various local problems. If we are able to rapidly respond to complaints from villages, this will demonstrate our efficiency and the speed of action that our project promises to the village. Thus the cooperation of authorities at the intermediate level can be very important to the project.

There is no question, either, of addressing ourselves directly to women before obtaining the total support of notables and influential men in the village. The real village Head is often not the most influential person. The religious leaders have frequently become people whom one must take into account. They are the people who can block the action of a project. It is necessary to listen to the caste men in the village [Editor's note: low caste singers or storytellers] who know the history of the village, its antagonisms, its intrigues, and how certain silences are to be exploited. It is necessary to know how to question the village elders and the young ones who effectively undertake the actions planned. After much coming and going, after meetings between our team and village groups to arouse enthusiasm and transmit information (women and men separately or together depending on the traditions of the milieu), we let them have time to think at length. We then chose nine villages in Kayes, six villages in Sokolo, and five villages in Dilly (the Dilly villages now wait for another source of financing).[2] The criteria for the choice of villages included those whose inhabitants accepted working with us, which were not isolated and had water, a large concentration of animals (cattle and small animals), a certain influence on other satellite villages, and a market. They also had to be villages in which there was much coming and going and a lot of interaction among sedentary people and sedentary and nomadic people. Once this work was done, it became necessary to choose village development agents.

The profile of the staff needed at the base was discussed during meetings with different sections of the OMBEVI/Pro-de-so after surveys in the field. These sections included the training and communications service, OMBEVI and Pro-de-so sociologists, the heads of the zones concerned, and those in charge of women's *animation* (such as it was then). During the following meetings, we helped women to organize themselves. They chose their own leaders taking into account criteria which we decided upon together. These criteria were that the leader had to be a woman between the ages of twenty-five and forty. She had to be a married woman or widowed, but not divorced, and a woman of high principles. She had to be living in the village and able to be available to help her sisters. The woman was to be chosen by the women villagers themselves and accepted by the village notables.

After choosing the pilot villages, and installing our woman leader and her contact committee (the staff responsible for the village *animation*), we tried to assist in the identification of their problems and analyze them together with the women to see how they have tried to solve them to date. The problems identified in most pilot villages in order of priority were: (1) a true need for water. Both women and men complained about wells which become dry too soon. Women pass most of their time at the still-usable water holes which are often quite far away. They have to get up too early (4 o'clock in the morning) and go to bed very late as a result. (2) A need for health care. Children were often victims of sudden fevers with rashes and all the associated nutritional illnesses (kwashior, etc.). After childbirth, women, as is often said in the village, have a foot in the tomb until the fortieth day after the birth. (3) There is also a need for early seeds (groundnuts, which is the basic food of our zones, cotton, gombo, baobab leaves, and monkey bread—are all needed). There is a need for condiments, too, with the eventual goal of having a condiment stock for current use.

After this first long period of making individual contacts, our group reported to the Board of Directors, which did not remain deaf to this information. (The chosen villages are mostly sedentary and composed of Soninke, Bambara, and Peulh. Maures are nomads and regularly move their animals, as do the Peulhs, along well known routes.) Before tackling the objective of our project—the *animation* of women, with the support of our administration—we had to find a solution to the water problem, for "without water, there is no life." The villagers were asking: Why do you prefer animals to people? Why do you prefer nomads to sedentary people?

Our project first dealt with pastoral drilling along the great animal routes. Our leadership, after much negotiation with donor agencies, started hydraulic work with the following results: In the Kayes-Nord region, 300 holes were drilled with 119 positive results. Fifty percent of these were village wells. In the Nara-Est of Sokolo region, there were sixteen positive drillings of which thirteen are close to villages; thirty wells or well-platforms were provided. The program of deepening ponds was maintained.

To address the wishes of the population and help them to equip themselves, certain health facilities have been inspected in the Kayes zone (Batama, Koniakary, Sero, Tiechibbe-Gory-Gopela, Tambacara, and soon

Tringa-Marena). Furthermore, a one-time distribution of seven tons of groundnut seeds (early variety 4710) was undertaken before the winter season of 1981, including the transportation of the seeds from Kita to Guidimakan. To solve the condiment problem, we are trying to obtain a small credit fund to provide the means to start a condiment stock. However, after carefully explaining the necessity to use these vegetables in daily food, seeds from the seed research center of Same-Kayes have already been distributed free of charge.

In our planning work in the field, during the first phase and through our field agents, we proceeded to gather socio-economic data in order to have a better knowledge of the physical characteristics of the village (geography, rainfall, pedology, streams, etc.). We also collected economic data on the village activities including agriculture, stock farming, rural industries, handicrafts, fishing, markets, exchanges, transactions, and, in the case of women's activities, listed the traditional activities of women in our zones. We took note of all traditional technologies as well.

We did a study of the socio-political milieu (What are the social strata? What is the traditional administration? Is it a gerontocracy? What are the relations between ethnic groups and between castes? We observed group dynamics, noted the traditional associations and the religions and made a sociogram during meetings). Studies were also made by BECIS (a research company) on the possibility of cattle breeders cooperatives, the role of women in the management of large herds and small cattle, and their roles in the processing of stock farming products in the Kayes-Nord zone. Reports were written by a Feminine Animation/FAO expert and her Malian counterpart on the role of women with regard to small animals, poultry farming, processing of dairy products towards a family industry, and milk marketing. The study of Dilly was undertaken by Marianne Rupp through a field survey in 1976 (see Rupp 1976).

In our surveys, we have particularly insisted on delineating the traditional role of women in the zones. It is in the sectors of production, preservation of agricultural products, stock farming, and handicrafts that women are the most involved. They are also responsible for child care. Women are nutritionists and manage the family budget. They take care of small gardens and poultry and they breed small and large animals. Besides, they remain responsible for water and wood gathering duty (a well-known scene of the African countryside and subject of so many postcards is a woman and her burden [water or wood] on her head and a small child on her back). Women provide from 60 to 80 percent of all rural labor, yet it is only since the Lusaka Conference in 1979 that women's work in rural communities is taken into account and that efforts have been made to lighten their tasks and increase their training to better integrate them in development.

Women of our zones have therefore appreciated the new efforts directed to them. "The support of the women of our zones has brought the support of their husbands even though women's action gives slower results and less spectacular ones." (Editor's note: No citation given.]

We have planned programs where women provide support during the execution and the follow-up of cattle vaccination and for sheep and goats in the areas of internal and external deparasitation, castration, and displacing the lambing post. We also have programs in forestry (the training for the construction of improved stoves) and in environmental preservation.

Results of Women's Animation 1980–83

The choice of 15 pilot villages was completed although activities in Dilly were suspended because AID financing was stopped.[3] The women's staff was put in place in the first villages. However, the second group of villages is late. The *animatrices* have been undergoing training for three months and they have just now gone back to Kayes.

The *animation* staff at the project level is frankly inadequate and we are waiting for new nominations.

We have formed a Contact Committee in each of the first villages including one woman to lighten tasks, one (low caste) woman for information and one (elderly) woman specializing in health and nutrition. There is also one woman for family economy.

Mobilization and "awakening" has been started and is 25 percent completed. Films on health (including clean milk, first aid care, and nutrition in Sahelian countries) have been shown. Visual aids have been acquired [such as posters and demonstration pictures and slides].

We are in the process of preparing, together with technicians from other services, an index card file of themes and the contents of training programs. This task is about one quarter completed.

We have made three donkeys and three carts available at Kayes and the same number at Sonkolo (although one donkey was eaten by hyenas). In the latter case the local population put up 20 percent of the purchasing price and are responsible for the use and management of the equipment.

Women use these carts to haul water and wood. Some women rent them to transport goods between villages.

Studies have been made in regard to the provision of grinding mills (for grains). These examine what mill is suitable given our feeding habits—the water-cooled Dental Super with Listen motor has been tested. This research is about 25 percent completed. (Before such mills are purchased, mechanics in agricultural equipment will be trained.)

Preparatory work in informing women about the construction of improved stoves has been begun before such stoves are actually built. We are employing the best national prototype with standardized norms [Editor's note: See Ki-Zerbo and Tucker on improved stoves].

The small garden project is in its second year. Women farmers have been very enthusiastic as they have been able to sell carrots at 50 FM each and still have some for home consumption.

The poultry farming program has begun and is 25 percent completed. A poultry agent has been trained while the choice of village vaccination agents and the replacement of local roosters by selected (thoroughbred) ones has been started [Editor's note: See Henderson on women's chicken project].

No village pharmacy has yet been established. Work on this program is about 10 percent completed.

Studies have included cattle, sheep and goat vaccination and castration, cattle fattening and the cleanliness, marketing, conservation and processing of milk. Programs will be developed when solutions have been found for all women's problems in these areas.

Searching for New Ways to Improve Animation Methods

Animation, which has always been directed from the top and tied to individual sectors in our developing countries, is taking on a new orientation. If the *animation* of the agricultural and handicraft world is much more advanced than the *animation* of cattle breeders and fishermen, nonetheless this latter *animation* is very important when one realizes the place that stock farming and fishing occupy in the Malian economy. Agricultural sector *animation* services are favored by research institutes which, together with commercial companies, work for the improvement of cash crops and have a hold on world market prices. However, one must not forget that the cattle breeding sector had a period of glory in history. The first animals brought to improve the stock came in Mali in 1896. During the Second World War, the stations of Fataladji (Segala/Kayes), El-Waladji (Macina), and Bamako were making butter, cheese, and dried meat which the Soudan, as it was then, sent to meatless regions of France. These efforts were plagued, however, with problems of animal epidemics and animal health.

Meat production in Mali was severely hurt by the drought of 1973 and spurred our Government toward efforts at *animation* in the area of animal production. *Animation* must now enable village women to identify and express their main problems and worries. Village women are concerned. It is necessary to go to them first as they are the ones involved, and it is essential to have this group express its concerns. They are the ones who know what they suffer from. Together they will find solutions. At least the fact that they are grouped together will force the authorities to take their complaints into consideration.

Suggestions for Animation and Training

Animation should begin with a preparatory meeting during which a questionnaire is distributed to arouse the interest and focus the attention of the village women. The questionnaire should ask what different activities are done in the village, what the major needs and problems are, and how the respondent thinks they might be solved. The budget of village activities should be studied to see how much income is generated by different ethnic groups and why. The different ethnic groups must be asked specific questions about what they do. The Peulh women can be questioned about how much milk they get, how much they sell or consume and in what form. Soninke women can be questioned about food crop cultivation and Haratin women (Black Maure) about leather work and tent weaving. All the women can be questioned about butter and handicrafts.

In addition to *animation* sessions, the village environment should be studied so that all the social organizations, economic activities, customs, and power structures can be understood.

The *animatrice* herself should be unassuming and able to speak to village people and participate in their activities—from working with the mortar to attending feasts, parties, and ceremonies around the well. An *animatrice* must be able to hold meetings for women in which she listens more than talks and, when she talks, uses analogies and language familiar to them. She must be well prepared and able to get most people to talk. She should be able to analyze where the problems come from—whether the cause is physical, economic, social or cultural—and be able to give examples of similar problems. When, finally, the problems and possible solutions are understood by everyone, she should be able to help the group find a solution and develop a plan of action.

Before this plan is undertaken, research will have to be done to find out how a similar plan has worked elsewhere.

The decision on what action to undertake must come from the local women themselves. The women must consider *who* will be responsible, what criteria will be used for their selection, and how they will be chosen. They will not be chosen by voting (which is not African), but by evaluating certain qualities which the profiles of different women will establish.

What action the village women wish to undertake must be clear and they must specify well-defined steps for its implementation. The women must decide *when* different things will be done according to a schedule which must be observable, measurable, and capable of modification. *How* things will be done—by what means—has to be decided. This will mean, perhaps, training, functional literacy for credit management, learning skills to maintain and use carts and grinding mills, etc. It will also mean finding financing.

At present we know that the success of a project depends on village women taking responsibility themselves and on the necessary training to help increase the productive capacity of the local women. *Animation*, which helps local women to think through and then make decisions themselves, will beget money, accumulated goods, and lead to self-sufficiency and the desired well-being.

Training is part of the process of developing know-how and knowing what to do. The trainer must know whom he is training and for what. Different village women will have different needs for training depending on what objectives they are trying to achieve. These objectives should be precise, pertinent, feasible, observable, and measurable. The problems blocking the achievement of these objectives should be clearly understood and then the necessary training established. *Animation* should be part of the training program which must be closely adapted to the local situation.

Two-way communication is very important in the training process because of all its effects on the knowledge, attitudes, and behavior of the trainee. Personal reactions will show whether the message has been received and if changes in the training program are necessary. Introduction of visual aids such as felt boards, figure series concerning local life, films, and video cassettes can have a real impact.

There are three classes of pedagogical objectives in training. The first is the field of knowledge which emphasizes comprehension. The second class of objectives is the field of psycho-motricity including the capacity to imitate, to connect things with each other, and to do things automatically without thinking. Finally, the field of effective impact, that is the change in attitude, which is more gradual and more difficult to evaluate, is the third area.

Training must start with the problems of rural women. There must be an interaction between what exists and the situation to be improved. Affirmative response and question techniques should be used. It is necessary to undertake the identification of the problem to be solved and help the trainee to discover what the trainer wants to teach him or her through a questionnaire. The trainer helps the trainee to discover what his own capabilities are through maieutic communication (the way Socrates brought out the thoughts of his students by forcing them to think). Therefore, training villagers allows them to face their difficulties better and their village structures will be more autonomous and independent of outside help.

Training may be undertaken through various types of sessions according to the training subjects—study trips, booklets, education series. The organization of a training session is like the preparation of a small training project. The following must be made clear:

> Why undertake the session?
>
> Who will be invited (whom to train)?
>
> Place, length, date (1–3 days).
>
> Logistic organization:
>
> Place of work, food, lodging, transportation, budget, and contribution of each trainee (in kind or in money).

The training session must then be prepared and the following tasks accomplished:

> Select the team of trainers for the session (2 or 3 people).
>
> Determine the role of these trainers.

Prepare the program and the different working sessions.

Decide the means of *animation* of the plenary sessions.

Plan an evaluation every evening and reprogramming if necessary.

Contact the technicians (if the themes are technical) and prepare with them the contents and methods of their lectures.

Invite regional correspondents, local authorities, administrative authorities, traditional authorities and government services.

At the start of a session, a word of welcome is necessary. Then it is necessary to break the ice through the use of amusing methods (for example, light a match, before it goes out the participant must have introduced himself). Then, the session: objectives must be introduced and also the themes, schedule, and practical organization. Eventually, responsibilities should be given to trainees (obtaining water, preparing meals, cleaning the lodgings). At the end of the session, conclusions should be drawn, the session evaluated, and the follow-up planned.

Conclusion

After experience in the field, certain major points emerge which are worth emphasizing here. In the first place, peoples' attitudes must be changed. Citizens must be made aware of their civic duties and they must be trained in those duties. People must receive information on the political options of the country at the regional, national, and international levels. The need to integrate women in development must be accepted.

It is important to return to old African principles. Africans are more social than technical, and these qualities must be used when working to change attitudes and behavior. Thus, traditional groups and associations must be made the base of local activities such as the *Tons* among the Bambara, the *karis* (men circumcised together at age 7), the *Flanbolo* among the Manding, and the *Naams* among people in Upper Volta and Madagascar.

Women must be made to feel concerned about rural development while men must accept the fact that their wives will undertake *animation* work. Development agents and *animateurs* must be retrained so that they work with the local farmer as their partner.

Secondly, all projects must start from a need felt at the base. *Animation* must be a central part of the project and competently done to prepare people to accept the projects and work with them. Training structures must be decentralized to facilitate animation, retraining, and mass training. Village credit schemes with village management must be started. All development schemes should be done in an integrated framework which sees the whole human being and makes him/her the center of the project.

A third point is the need to improve women's training in particular. Women's training is insufficient at all levels and the staff is inadequate. Women who are part of the training cadres are not properly motivated. Bonuses, scholarships, and proper logistic support would help improve motivation. So, too, would study trips and exchanges.

The fourth point is the lack of economic means available to women. They lack access to credit, so credit systems must be set up adapted to their small-scale local activities. In setting these up, their traditional social groups should be taken into account. Furthermore, a heavy administrative structure should not be imposed on them as they are either illiterate or scarcely literate.

Finally, let me say, someone for whom communication is very difficult will not be a good *animatrice*. The *animatrice* must be unpretentious with an easy manner and good at working with people. She must be enthusiastic about helping others to open up, and she must favor participation by local women as much as possible. She must support and facilitate local responsibility, thinking, searching for solutions, and decision-making. Vincent Cosmao said, "Development is conscious and mastered action, collective and global, of a responsible community to undertake its mode of life so as to realize the human project it chooses." [Editor's Note: No citation provided.]

This, in a few words, is the experience I wanted to share with you. There can never be enough insistence on *animation* for integrated development projects which are trying to help people find solutions to their problems both within their groups and as a whole.

Notes

1. *Editor's Note:* This may be an optimistic assessment although Mali may develop as a major supplier of meat in the future.
2. *Editor's Note:* With USAID's withdrawal from funding, the Dilly villages no longer have support for programs planned by Mme. Kantara's group.
3. USAID, which funded the women's program, not the larger stock-farming project, apparently did not see sufficient progress in the program to renew its funding after the first grant (Interviews 1985).

7 The Grassroots Women's Committee as a Development Strategy in an Upper Volta Village[1]

Helen Henderson

Editor's Note

Dr. Helen Henderson is an anthropologist who succeeded Dr. Cloud as the director of the Woman's Food and Information Network. She is presently attached to the Bureau of Applied Research in Anthropology at the University of Arizona.

Dr. Henderson's chapter adds an important dimension to this book. She describes the patterns of interaction of the women of the ethnic groups and their agricultural practices in the area of Upper Volta (Burkina Faso) where she worked. This provides an addition to the material in the first section of this book on Mali and Senegal and substantiates the view that women have different roles in agriculture, depending on their ethnic group, age, and family status, and on the availability of resources (among other things).

Dr. Henderson was attached to a project which carried out research on women in the area of a large stock farming program in 1978–1979 to find out what strategies would most usefully increase their productivity in the area of livestock management. At the conclusion of her work, USAID did not fund the second stage of the program. This decision had nothing to do with the findings of her project team, but rather it had to do with problems AID had with the larger project, the funding of which was also not renewed. It is interesting, and also a negative commentary on many projects, how the women in Dr. Henderson's area waited in vain for help they had come to expect, and how the development efforts they did carry out on their own failed.

The Setting

For a period of three months in 1978–1979, my research assistant and I took part in an ongoing livestock project in an Upper Volta village in order to assess the social and ecological environment in which rural women live and to elicit their participation in designing further development projects that might be desirable and useful to them.[2] Our intention was not to "sell" externally chosen projects. Rather we wanted to encourage grassroots women's groups or committees to become involved in the process of development. Thus they would shape the project with their understanding of current livestock activities of village women and knowledge of seasonal and daily time constraints affecting potential expansion of women's roles in livestock production. We also hoped to delineate the current and possible organizational bases for cooperation among village women, and identify female opinion leaders within the different ethnic groups of the village.

Being part of a livestock project associated with the government livestock agency for Upper Volta, we were, of course, constrained to focus our attention on livestock issues. We did not intend to give advice on nutrition, gardening or infant health, though such issues might be of primary concern to local women.

At the time of our project, the ideas of grassroots committees and "development from below" were popular in some quarters of major funding agencies. The implications of such ideas, however, had not been carefully examined. The grassroots women's committee in this particular village had, for its model, the already established village livestockmen's committee, which consisted of men from the three major ethnic groups and the two major religions in the village. Establishment of the livestockmen's committee and subsequent construction of a highly desired cattle vaccinaton park had been early achievements of the project and had given the villagers positive expectations toward project personnel, paving the way for acceptance by the men of our research among women.

Our methodology involved systematic observation of village women in their daily routines, interviews with women on selected livestock-related topics, meetings held with groups of women to ascertain their opinions concerning livestock practices, and development and administration of a questionnaire which was given to seventy-one women concerning their current livestock practices and their attitudes about possible women's roles in development of the livestock sector.[3]

Within the village of Koukoundi semi-sedentary Fulani herders live alongside Riimaaybe farmers (formerly their slaves) and Mossi agriculturalists. The Mossi are further subdivided into Mossi animists (followers of the traditional religion) and Mossi Muslims. Members of each group possess distinctive organizational bases and distinctive attitudes toward cooperative work both among themselves and with other groups. In the following sections we will briefly sketch the major socio-economic features of women's lives and the differing orientations toward grassroots organizing which we observed among these diverse groups from this one village in Upper Volta.

The village of Koukoundi is located approximately 90 kilometers north of Ouagadougou (the national capital). Its mixed ethnic population consists of approximately 63.2 percent Mossi and 36.8 percent Fulani and Riimaaybe in a total estimated population of 504. The number of adult females is estimated to be 150. The village has no school (other than Quranic) and no dispensary. Its weekly market is small and was established only a few years ago. Although government agents visit the village with some regularity, they do not usually interact with the females but, according to the women, "speak only to men."

Description of Ethnic Groups

Fulani Women

The village of Koukoundi was settled more than eighty years ago by the ancestors of the current Fulani chief (Vengroff 1980, 9). When the Mossi later settled in the area, they consulted this chief for certain administrative needs, but went to the Mossi chief of a neighboring village for traditional matters. Neither group considers itself subordinate to the other, and each practices a high degree of public cultural tolerance. Most Fulani speak More, the Mossi language, though most Mossi do not know Fulfulde, the Fulani language.

The Fulani are organized into patrilineages, within which moral consensus is a stronger factor in sociopolitical control than is formal authority. In general, Fulani are highly individualistic and tend not to think in terms of community efforts. According to Riesman, cooperation for the common good hardly exists, though reciprocal exchange of help on a one-to-one basis is common (Reisman 1977, 73). The patrilineage does not provide a strongly hierarchical framework for men or for women

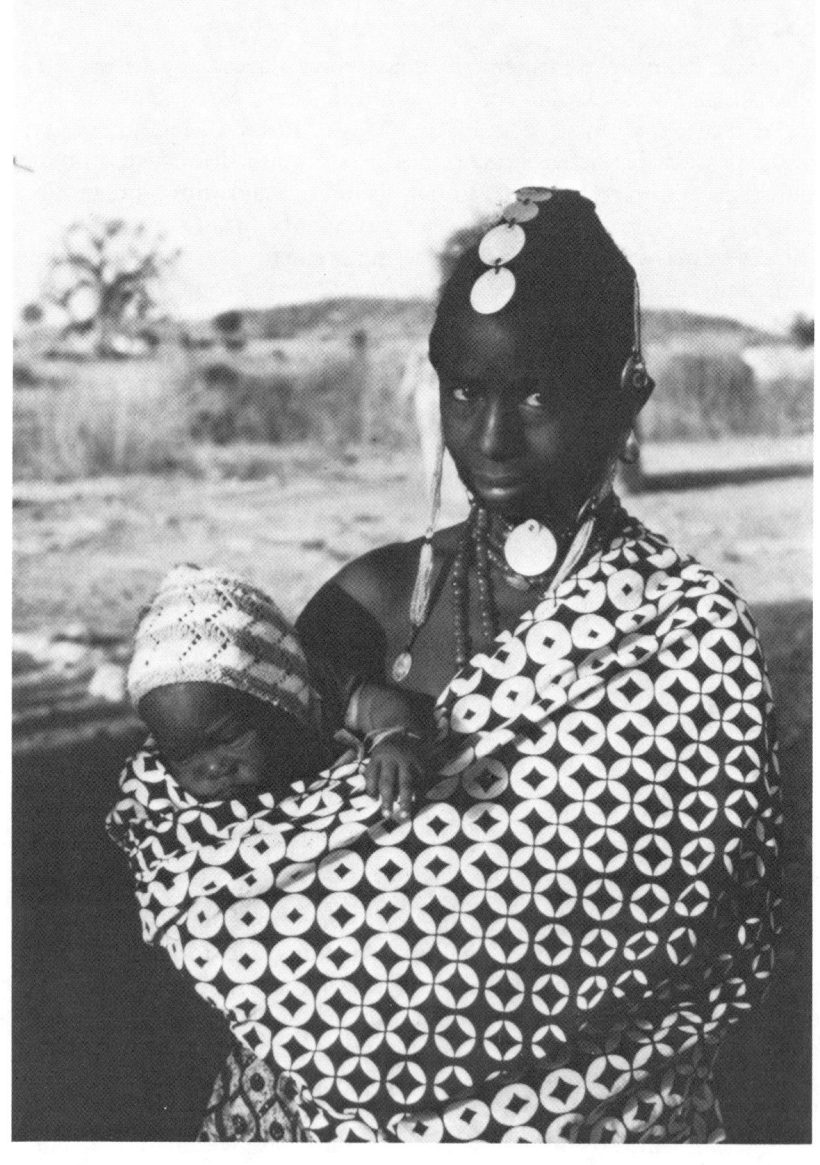

A young Fulani woman and her baby in the village of Koukoundi in Burkina Faso. "Just as a [Fulani] man's authority is primarily based on his personal characteristics, the influence a woman can have on the community can not be related to her social structural position (birth order or prestige or family) but depends entirely on her personality. Women tend to work by themselves, helped by their children" (Reisman 1977). Photo by Helen Henderson because there is no formal "head of women" who has authority over the other women of the patrilineal segment. Even among wives, the senior lacks authority and power over the junior. There are few everyday activities that bring the larger group together, although individual visiting is common. Kin-relatedness is emphasized among Fulani women of Koukoundi and visiting of relatives is a popular pastime. Males of the same patrilineage share vaguely defined territories, herd cattle together at times, and tend to live in minimal clusters near closely related lineage members.

Just as a man's authority is primarily based on his personal characteristics, the influence a woman has on the community cannot be related to her social structural position (birth order or prestige of family) but depends entirely on her personality (Reisman 1977, 57). Women tend to work by themselves, helped by their children. Small clusters of women may, however, often be seen pounding millet together.

All Fulani in Koukoundi are Muslim, as are almost all Riimaaybe and approximately half of the Mossi. Among the Fulani and Riimaaybe, Islamic rites such as naming and marriage occasionally draw large numbers of women and men together. Annual Islamic ceremonies (Ramadan and Tabaski) unite the Muslim men of all three ethnic groups, but Fulani women pray individually, even at these times.

Fulani women in Koukoundi are not heavily engaged in agricultural tasks. About one-third of our sample of twenty-four women indicated that they do not farm, and only a few said they grow millet, beans or cotton. During the rainy season okra, *oseilles,* and pepper are commonly grown in gardens. Fulani women's primary agricultural labor output involves assisting in the millet harvest by carrying already cut millet heads to the compound. They tend to assist only their husbands, not members of the extended family. None of the women interviewed say they sell any agricultural produce.

More than one-third of the Fulani women sampled said that they currently own cattle. They are, however, reluctant to talk about their rights in cattle, especially when other Fulani women are present. Rights in cattle are closely articulated with those of other family members. A Fulani woman in Koukoundi is considered relatively rich in cattle if she has six. Most of the women who told us they own cattle reported having received them from their own relatives and none purchased them themselves.

Before selling a cow, a woman must have the permission of the man in whose herd it is kept, and she may only sell through a man. Prior to her marriage, the major reasons given for sale of a woman's cattle are the need to purchase jewelry. After marriage, women rarely sell cattle, preferring to keep them for their sons. Those who reported having sold them, however, did so to pay for their own medical expenses or for millet purchases in times of famine.

Young women go on transhumance with their husbands, but in general are not responsible for herding cattle or taking them to waterholes. However, women sometimes go ahead, to look for new pastures. Vaccinations and other medications for animals are usually paid for by the men in whose herds the animals graze.

The importance of cattle to Fulani women lies particularly in the milk they produce. All of the women who were interviewed milked cows. As women enter their husbands' households, they are allocated cows for milking purposes. Women may also milk cattle belonging to their children, as long as the animals remain in the family herd and are not redistributed to the son's wife or removed by a daughter to her husband's herds.

All of the Fulani women we interviewed reserve part of each day's milk supply for their family's use. However, in the rainy season, when there is a good supply of milk, more milk is sold than is kept in the household. As the dry season progresses, cows produce less milk, and in January a woman may obtain only one liter from three cows, compared with six liters during the rainy season (Henderson 1980: 123). When there is sufficient milk supply, Fulani women also make soap, butter, and yogurt.

Milk clearly is the major source of disposable income for Fulani women of Koukoundi. Their major trading clients are the Mossi and the Riimaaybe, who purchase milk in small scoops for 5–25 CFA to give to their children. Proceeds from the sale of milk are used to buy jewelry, condiments, cloth, and millet (Henderson 1980, 123).

During the season of plentiful milk, the Fulani women make daily morning trips to the Riimaaybe compounds and the dispersed households of the Mossi. Three hours is considered a brief amount of time to spend in this activity. Although visits to the Mossi compounds are primarily economic, social greetings are exchanged and news of family members communicated. Although they are not direct participants in the web of Mossi social activities, Fulani women act as casual information channels among the various ethnic groups in the village. In the weekly village market, small groups of Fulani women sit together selling milk and dairy products.

Riimaaybe Women

Riimaaybe are descendants of people captured by the Fulani and, in former times, they cultivated fields for their masters. The Riimaaybe of

Koukoundi live in a settlement cluster near the Fulani chief but also not far from neighboring Mossi compounds. Riimaaybe use the same family names as the Fulani and are to some degree attached to the latter's patrilineage segments. Although Fulani men and women do not marry Riimaaybe, some of the Riimaaybe in the village have Fulani genitors. In religious matters, Riimaaybe now participate in the same Islamic sect as the Fulani of Koukoundi.

Riimaaybe women participate in Fulani patrilineage and other social events. They will, for example, dance at a festival to celebrate a Muslim boy's literacy in the Koran. Young Fulani and Riimaaybe women may be frequently seen in the Riimaaybe settlement pounding millet together, laughing and joking. It is much more unusual to see a Riimaaybe woman in the Fulani compound, unless she is there simply to give greetings or perform some menial task (for which she is paid nowadays). However, although some elderly Riimaaybe men show subservience to elderly Fulani men, younger Riimaaybe do not express such attitudes. They demand to be treated with respect and paid for their work. In general, Fulani appear to feel more at ease in Riimaaybe compounds than their own. Some of their strict rules of decorum are set aside temporarily in Rimaaybe compounds (Cf. Reisman 1977, 121).

Riimaaybe women do planting, cultivation and harvest work. Almost all the women we interviewed say that on compound land they farm sorghum, beans, cotton, corn, and millet. In their own fields or gardens, they grow okra, peanuts, sorghum, sesame, and beans. All report that they help their husbands and others with their millet harvest and are in turn helped by men and children of the compound.

Though they farm, they sell little of the unprocessed produce. Instead, they conserve it for family use. Most Riimaaybe women say they do not have their own personal granaries. The major unprocessed crop sold is peanuts. The primary marketing activity of Riimaaybe women, however, is the selling of millet flour and millet flour balls. Proceeds are used to buy jewelry, condiments, milk, and kola nuts. Some of the Riimaaybe attend market on a regular basis, often sitting with the Fulani women.

A few Riimaaybe women own cattle, and women whose husbands have cattle, milk them as do Fulani women. None of the Riimaaybe women, however, sell milk. Indeed, 81 percent of the 11 women interviewed said that they regularly buy milk from the Fulani. When there is sufficient milk, women also make butter and soap, but usually not for sale. Approximately half of the Riimaaybe women interviewed say that they would advise a woman to purchase cattle as an investment because of the milk.

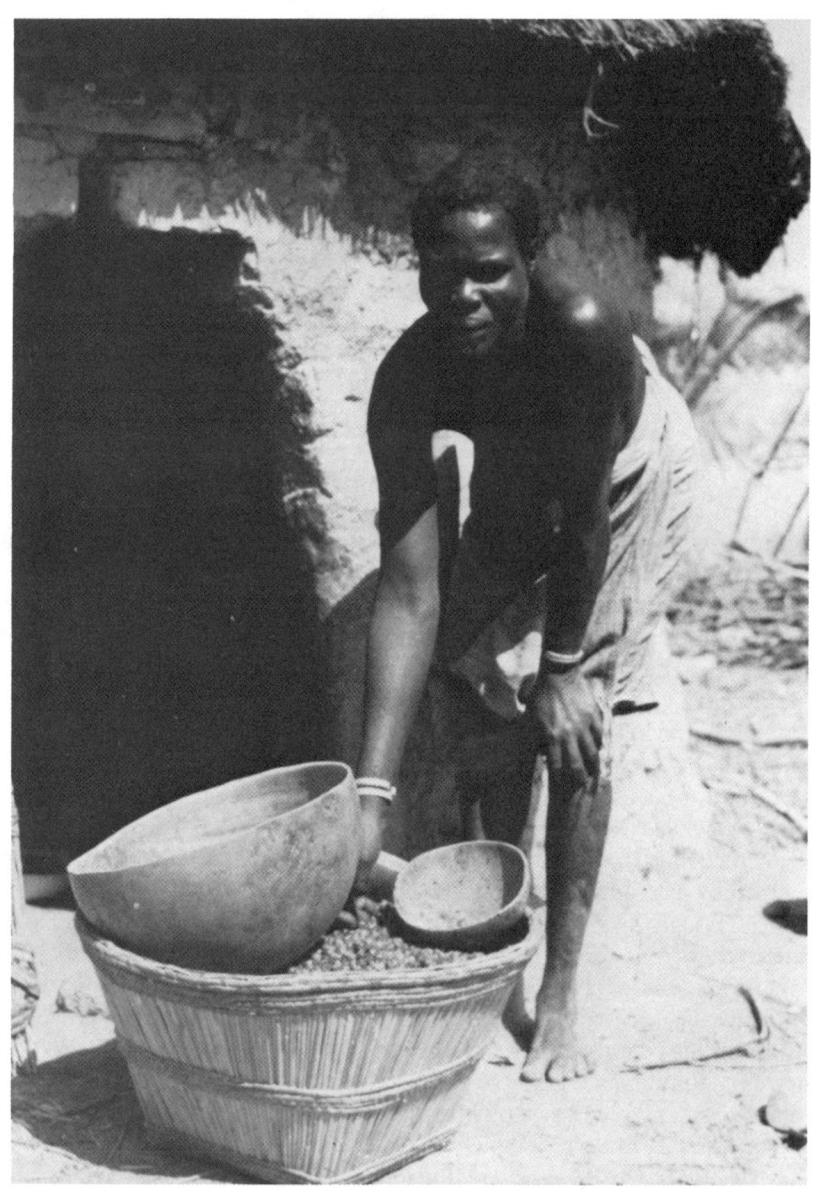

Mossi woman with beans prepared for a work party harvesting cotton. Mossi women are engaged in agricultural labor throughout the farming season, working not only on communal fields but also on land allocated to them by the head of the family for their own agricultural production. Photo by Helen Henderson

Mossi Women

Historically, the Mossi have been members of a centralized and hierarchical kingdom. Land is held collectively by male members of the patrilineage. The patriclan forms a residential kin group with dispersed compounds comprised of polygamous extended families in which husbands and wives have individual huts (Hammond 1966, 110). Joint farming activities are extremely important.

Within a compound and among neighboring patrilineally related compounds, Mossi wives form a tight communication network. As a collective group, the lineage wives can be mobilized by the head wife of the lineage elder (Skinner 1964 refers to the *pughtiema*). Although the *pughtiema's* major decisions must be authorized by the lineage elders, she wields considerable influence both in organizing women's work (for example food preparation at ceremonial occasions and women's exchange labor parties) and in settling disputes among wives. She is often an advisor on matters of childbirth and other situations common to women.

Patrilineally related groups often form the basis for neighborhood organization among the Mossi, especially if these groups share religious affiliations.

Currently, approximately half the Mossi in Koukoundi are Islamic, the number having increased steadily during the past generation. Muslim Mossi are set apart from animist women by their religious observances, including prohibitions against making millet beer and against eating certain foods. Islam unites Mossi women from distant or unrelated compounds, but it does not unite women from different ethnic groups (Fulani, Riimaaybe and Muslim Mossi women). Mossi animist and Mossi Muslim women do come together for common activities on some occasions, such as funerals. At annual festivals such as Ramadan and Tabaski, some of the Mossi Muslim women participate in public prayer groups and also go from compound to compound saluting Muslims, including the Fulani.

Mossi hold certain negative stereotypes about the Fulani, especially Fulani women, who are not cultivators and therefore unacceptable as Mossi wives (Skinner 1964: 210). Mossi women speak of the Fulani women as "too proud," "too noisy," and "not working enough."

Mossi women are engaged in agricultural labor throughout the farming season, working not only on communal fields but also on land allocated to them by the head of the family for their own agricultural production. Almost all the women we interviewed hold such land. Almost all grow white millet, corn, beans, sorghum, peanuts and cotton. The agricultural produce from the women's fields not only helps feed their families, but gives the women some disposable income. The majority of Mossi women have their own millet granaries, in sharp contrast to Riimaaybe and Fulani women (Henderson 1980: 133).

Husbands assist their wives in difficult farming tasks, such as clearing of land and harvest. Women reported that, if a large work party is needed during harvest, they can request the help of other women outside the immediate compound, using the head wife, or *pughtiema*, as their representative.

Mossi women assist their men in the communal fields and their personal fields. When asked who they helped in the millet harvest, almost all the 34 women interviewed named their husbands, and about two-thirds cited other women in the household. It appears that women spend more time in their husband's fields now than they did in the past, when men's fields were smaller and the family labor group was larger (see Hemmings N.D., 26).

The major source of disposable income for Mossi women is sale of agricultural produce (both raw or in processed form). Peanuts, millet, and beans are the major crops sold, usually in the village market. A few women in our sample sell to village traders, travelling merchants, or agents of the government. We are told that Mossi women often travel in small groups to nearby markets in neighboring towns, where they can get slightly higher prices for their goods. Frequency of visits to these markets has declined, however, since the establishment of the Koukoundi market. In the market, Mossi women sit together in stalls, along with others of their husband's patrilineal group or their neighborhood, selling similar agricultural products. Aside from crops they have from their own or their husbands' fields, they also sell prepared foods, dried fish, fruits, eggs (collected in the bush), kola, and many other local items. Women indicated that they bought milk (from the Fulani), kola nuts, clothing, tobacco, jewelry, and animals with the money earned.

Although some of the Mossi men own cattle, none of the Mossi women interviewed reported milking cows or owning any cattle. Slightly over half of the women frequently purchase fresh milk from the Fulani, and twenty-one of the thirty-four Mossi women interviewed buy this milk with money earned from the sale of agricultural products. Somewhat surprisingly, 38.2 percent of the Mossi women interviewed said that cattle would be the best livestock investment for a woman because of the milk. The dairying activities of Fulani women may be models for Mossi women; they evidently are models for Riimaaybe women.

Economic Activities Common to Women in the Three Groups

Women in the village also keep goats, sheep, and poultry, as can be seen in Table 2. Poultry and goats are the most common animals sold. Among the various ethnic groups, a higher percentage of Riimaaybe women claim to own goats and sheep. However approximately one-third of the Mossi women interviewed own goats, and three-fourths of those had sold goats the previous year. Proportionately more middle-aged (40–49) women own goats than do those in other age categories. Although small ruminants are sometimes received as gifts, nine to twenty-three

Table 2

Women's Livestock Ownership and Sale

	Mossi N=34	Fulani N=24	Riimaaybe N=11	Other N=2	Row Total N=71
Own livestock	58.8% (20)	66.7% (16)	90.9% (10)	100% (2)	67.6% (48)
Own goats	35.3% (12)	20.8% (5)	45.5% (5)	50% (1)	32.4% (23)
sold goats last 2 years	23.5% (8)	4.2% (1)	9.1% (1)	50% (1)	15.5% (11)
Own cattle	0%	29.2% (7)	9.1% (1)	0%	11.3% (8)
sold cattle last 2 years	0%	8.3% (2)	0%	0%	2.8% (2)
Own sheep	5.9% (2)	4.2% (1)	18.1% (2)	0%	7% (5)
sold sheep last 2 years	2.9% (1)	0%	9.1% (1)	0%	2.8% (2)
Own poultry	58.8% (20)	41.0% (10)	81.8% (9)	100% (1)	57.7% (41)
sold poultry last 2 years	38.2% (13)	33.3% (8)	45.5% (5)	50% (1)	38% (27)

women in the total sample purchased animals. In our sample, Fulani rank third proportionately in ownership of small ruminants.

All the women regarded small ruminants as an investment and Mossi women emphasized that these animals are a hedge against famine during the beginning of the rainy season. They added, however, that if a woman sells an animal to buy millet she is expected to share the food with others of her family, including co-wives and their children. All the women said they need their husband's permission before arranging to sell livestock, and none have personally made such a sale. Instead, all utilize males as their agents.

Women care for small ruminants in the compound, but the animals are herded by village youths. Women expressed concern that so many of their animals are dying of diseases, but they are not well informed on possible preventative measures.

One half of the women interviewed said they own chickens, but the proportion of Riimaaybe claiming to own chickens is almost twice that of the Fulani. Women between ages 30–49 make up the majority of those who own poultry. About three-fourths of the female poultry owners said they purchased the animals themselves, compared to nine of the 23 goat owners. Most women keep poultry around the compound and sell their chickens themselves. Women also complain about high losses of chickens to disease.

Domestic Activities of Women

In all three ethnic groups, women spend many hours a day processing and preparing food and collecting water. They also gather firewood several times a week. In families where there are several adult females (especially co-wives), women may take turns preparing the main meal for the household. The major difference in food preparation is that, while Fulani and Riimaaybe women pound millet in a mortar to process it into flour, the Mossi women grind it on a waist-high, circular platform situated in a central area of the compound. In the evening it is not unusual to see ten or more Mossi women processing millet at the grinding platform. Fulani food processing requires fewer women at the grinding platform. Young married Fulani women often go to the nearby Riimaaybe compound to prepare millet with Riimaaybe friends. It is more unusual for Riimaaybe women to come to the Fulani area to share such a task.

Women in all three groups spin cotton thread by hand in the dry season. However, we observed many more Mossi than Fulani women making thread. While Mossi women in Koukoundi usually wear local cloth (spun and then woven by men of the village), Fulani women frequently wear imported cloth.

Execution of the Project

Identification of Women's Networks

During our first week in Koukoundi, we requested the Chief to summon the adult women of the community to meet in the market area and hear about our research. More than eighty Fulani, Riimaaybe, and Mossi women appeared. We briefly explained our interest in obtaining data concerning local women's activities with livestock, their problems in this area, and their hopes for the future. We emphasized that the Livestock Service was already working with the men in the village and now wished to expand its activities to include women. We explained our methods for inquiring about women's busy lives, and then asked for their opinions. Repeatedly, we were told that while women are interested in keeping livestock, their major problems are that their animals die, and that they lack the resources with which to replace them.

In our subsequent meetings with indigenous groups of women, we tried to note, where possible, the compound and the religion of the women attending. We also noted what topics they talked about and where they sat. After each meeting, with the help of a census list and a map, we were able to locate the women's compounds in relation to others who had attended or not attended the meeting. In this manner it was possible to systematically observe common patterns of female interaction.

Although our basic methodology had been decided prior to meetings, we made two important changes in our approach, in response to demands made to us by a group of Mossi Muslim women. The spokesperson for these women said that although they understood that we were here to study and ask questions, they wanted to know if there was anything directly and immediately helpful that we could do for them and their children. We then decided to make some veterinary information available to the women (with the cooperation of the project's male veterinary extension worker) and to provide, when desired, some demonstrations of the use of locally available supplementary weaning food, including fresh milk, millet, and peanuts. Our recipes for use of these foods was modelled on those used elsewhere in the Sahel (Belloncle 1975).

Meetings with Individual Ethnic Groups

FULANI

A total of four meetings were held with groups of Fulani women to discuss their interests in livestock, to elicit their project suggestions, and to provide the nutrition demonstration. All but one meeting was at our initiative. Attendance at the meetings varied from thirteen participants to five. All meetings were held at the houses of middle-aged to elderly women who had been identified by other Fulani as being prominent by reason of activities, personal wealth, and status of husband or son. At all meetings the women stated that they had no cattle, few goats and that most of their chickens had died. Apparently none were familiar with medicines or vaccines available for their animals, though some expressed an interest in learning about them. When we discussed income-generating projects, some groups of women said they wanted to keep more livestock, but others said that Fulani women in this region did not work much now and did not want to work more in the future. They said their hardest work was selling milk, carrying water, and pounding millet. No spokesperson emerged in these groups, and a number of women tended to talk at the same time. One meeting was scheduled but cancelled for lack of attendance.

The problems encountered with these meetings are indicative of wider difficulties we experienced in identifying effective leaders among Fulani women. There are few structural guidelines for locating female Fulani leaders. It is probable that Fulani women would be more interested in discussing projects involving milk production than those concerned with animal health. Their animals are in the herds of their husbands or fathers, and these men are expected to provide vaccines. Women do not deal with cattle directly, and, in fact, rarely sell or buy them. Nor did women indicate that they sell goats or sheep frequently. Women do keep poultry in the compound, but several elderly Fulani women indicated to us that giving extensive attention to chickens would be inappropriate behavior for Fulani women.

It is also apparent that during the dry season, when milk sales are low, Fulani women have considerable leisure time for visiting neighbors in late morning and early afternoon. However, difficulties in obtaining water for additional animals in the dry season and heavy work schedules for women in the rainy season could present problems for the successful operation of a poultry project. Furthermore, the fact that women have leisure time does not mean that they wish to take up new cooperative economic activities.

RIIMAAYBE

From observations made at the initial community-wide meeting and from conversations with women in the Riimaaybe neighborhood, we identified as a possible leader an elderly kola trader whose husband owned cattle and was active in the project's livestockmen's committee. Upon her invitation, nearly all Riimaaybe households were represented at the meeting. The high level of participation reflected not only the women's interest but also the residential proximity of Riimaaybe compounds to one another. A few young Fulani women from the area also attended.

Although there were clear differences in wealth and status among the members of the group, even those women who were not economically well-off appeared to feel free to participate actively and voice their opinions. The Riimaaybe women were very concerned about the serious health problems of their animals, especially goats and poultry. Several women said almost all of their goats had died the previous year. When we brought up the issue, they showed interest in the possibility of poultry vaccinations and volunteered to pay the cost of immunization. When they were not in the presence of Fulani women, Riimaaybe told us that they would be interested in a poultry project but did not want to work with the Fulani, who were "too proud" and would try to take command.

Problems of human health were, however, more pressing to the Riimaaybe women. They spoke of the absence of medical facilities, the inadequacy of medical supplies in nearby towns and the lack of any vaccination program for humans in the village. They were deeply concerned about their children's illnesses and their own problems of difficult childbirths.

MOSSI

Of the three ethnic groups in the village, the Mossi were the most persistently interested in our project activities and had the highest attendance at our meetings. We held four meetings and nutrition demonstrations with the Mossi animists and attendance ranged from 55 to 25. All groups of women spoke repeatedly about their losses of goats and chickens. When a group expressed interest in having their goats vaccinated, we brought the project's local extension agent to provide them with more information. Two of the groups said they would be able to pay for poultry vaccinations.

Even in the one patrilineal group where women told us they were not allowed to own animals, there was interest in the possibility of a poultry project which would be sanctioned by the government and thus persuasive to the men. An elder spokeswoman said, moreover, that projects run by women would protect the women's financial interests better than those also involving men.

Initially, all of the animist groups were slow to organize themselves. Though they wanted nutrition demonstrations and animal health information, they were more keenly interested in infant health care, which we were not able to provide. They pointed out that if women were in better health, they would have more strength to participate in livestock programs. In all animist groups, the major leaders were the elderly, traditional ones.

The best organized and most consistent group participating in grassroots meetings in Koukoundi consisted of 20–35 women representing numerous Mossi Muslim compounds. This group sought us out numerous times to get information they hoped would be useful to them. During the meetings, women discussed the deaths of their animals and the general needs of the community. Upon request, meetings were also held to discuss not only livestock matters but also first aid measures and the possibilities of communal gardening.

The concerns of the Mossi Muslims differed from those of the animists mainly in the effectiveness of their organization. There were two leaders, one the traditional, elderly *pughtiema,* and the other a woman whose husband worked for the Autorité des Amenagements des Vallées de Volta (AVV Project), thus providing a possible contact with development work outside the village.

Before leaving Koukoundi, we asked each of the major groups with whom we had met in separate grassroots committees, to come together to form a central committee that would represent the livestock interests of women in the village to the livestock project and representatives of the National Livestock Service. A meeting was held consisting of representatives of each of the major Mossi groups (Muslim and Animist), the Riimaaybe and the Fulani, at which the women restated their problems concerning livestock and their hopes for the future. As a major concern, they expressed the need for water if there was to be increased livestock activity. They also spoke about the need for not only medicines for animals, but also for better medical facilities for human beings. The women affirmed that they would be willing to undertake a poultry project, and said they would continue to meet and discuss issues with local personnel of the livestock project.

After our departure, and unanticipated by project personnel, the group of Mossi Muslim women and a few animist women previously associated with them again expressed interest in a poultry project. On their own initiative, they organized themselves into a cooperative based on local work patterns and neighborhood, ethnic, and religious loyalties. This "cooperative" existed prior to the project's arrival in the village, and its core was made up of approximately the same group of Mossi Muslim women who had previously requested nutritional and health information. However, now they had become a grassroots women's development committee.

The grassroots women's committee established a small poultry project with the help of their husbands, bringing over 60 chickens to the project's local vaccinator to be immunized against certain poultry diseases. Over a six month period, the women's committee continued to meet with visiting project staff and local livestock extension agents.

The same group of women also planned to begin a communal garden with the hope of earning enough money from the sale of cash crops to buy medicines and other improvements for the village. They had been encouraged to undertake such an activity by the visit of a Mossi female extension agent from a neighboring area.

It should be noted that the Fulani were not involved in this particular indigenous grassroots committee. As already indicated, they are not primarily concerned with poultry or with gardening. Undoubtedly, stereotypes held by both groups impede egalitarian social interaction even when economic interests, e.g. in poultry, are shared. Riimaaybe women, however, do share many of the interests of the Mossi Muslims in matters of poultry and small ruminant management. Their holding of livestock appear to be significant and their interest in development activities high, as expressed in their own grassroots meetings. Yet they were not participants in the committee, probably due to their lack of social and economic bonds with the Mossi organizing group. Ironically, although the Riimaaybe women are thoroughly accustomed to working with Fulani women, they fear being dominated by them.

At present, the Koukoundi grassroots women's committee is at a standstill. Although the women had their chickens vaccinated, and mobilized their husbands to build a poultry breeding unit with local materials, they waited months for the promised delivery of an improved breed of

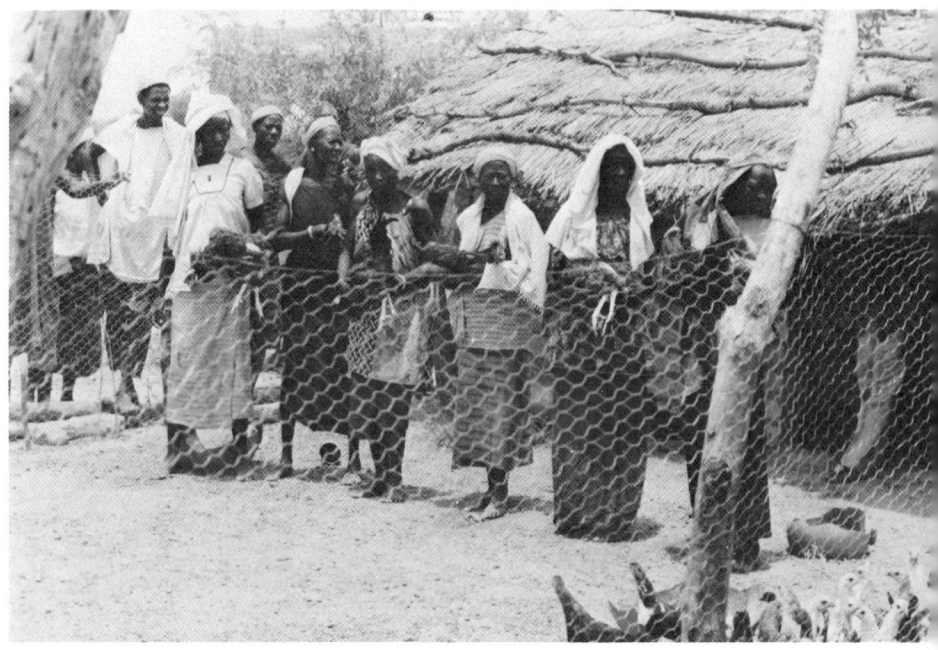

These women of the village of Koukoundi organized themselves into a cooperative and began a project to improve and expand their poultry farming. As part of their project, they were to receive an improved breed of roosters. They waited for months for these to arrive and are shown in this picture on the day the roosters finally were brought to the village. But the breed may not have been well chosen by the donor agency for they soon died, leaving the women disappointed. Photo by Helen Henderson

roosters—which died a few months after their arrival. A well, which the women considered very important, was begun by the project but never finished. Project activity ceased when Phase I funding ended. The male livestockmen's committee is in a similar situation, although there is some prospect that limited support for some of their activities will continue through the government livestock agency.

Phase I of the Upper Volta livestock project is now ended. It was devoted primarily to providing baseline data and building a foundation on which to begin Phase II (the major implementation activities). Its goals were primarily to identify local leaders, local problems, and local solutions, using grassroots or "development from below approaches." However, prior to the conclusion of Phase I, the funding agency decided not to fund Phase II but rather to focus its endeavors in other regions and on other problems. To what degree the national or regional livestock service of Upper Volta will continue to take an interest in these small, locally initiated projects, of moderate economic impact and serving the development of an out-of-the-way village, is unknown at present.

Conclusion

International funding agencies and eager project personnel often forget that their very action of inquiring and conducting base-line research into community needs may stimulate sectors of the community to establish development efforts of their own. These local people are bound to believe that the presence of official outsiders (foreigners or nationals) is a sign that further positive action will be forthcoming. It is difficult—perhaps impossible—to explain to them that they are a "sample of a wider population" with which the project may later intend to deal.

I suggest that whenever a village is used as a research population, project personnel and agencies should also make a commitment to work with local people to design and carry out some local improvement project within a reasonable time span. I also suggest that it be written into any development contract that when a long-term project is cancelled by a funding agency, both representatives of that funding agency and project personnel will return to the research population and explain what has happened and how they now intend to handle those issues that they have raised but left unfinished. If, as in the present example, local committees, were encouraged when this was a fashionable development strategy, how will these committees now be articulated with other agencies of the region or what new avenue can be provided so that local opinion can be heard? Accountability to local populations must be an integral part of project design and implementation.

Notes

1. As indicated at the beginning, the name Upper Volta has been retained although, since this chapter was written, the country has officially become Burkina Faso.

2. Research was conducted under the auspices of the Upper Volta Village Livestock Project. I want to thank Madame B. Balima-Zorgrana, Dr. W. Gerald Matlock, Dr. R. Vengroff, Dr. R. Henderson, Dr. D. Cleveland, Mr. M. Sourabie,, Ms. J. Heagan, and Mrs. Evelyn Jorgensen.

3. For more detailed information on the survey material see Vengroff (1980) and Consortium for International Development (1980).

8 Activities of the Women's Promotion Division of the National Functional Literacy and Applied Linguistics Board (DNAFLA-DPF)

Dembele Sata Djire

Editor's Note

The value of non-formal education programs has been debated both in theoretical writings and when actual programs have been evaluated. Non-formal education, and in particular functional literacy programs, have the advantage over formal schooling of reaching people who would not otherwise be educated, such as illiterate adults (especially women). Functional literacy programs teach basic vocabulary, reading, writing, and simple arithmetic in the indigenous language. In a short period of time, those who attend should be able to use these skills in everyday activities. For example, they should be able to keep simple records of their financial transactions or read instructions on fertilizer sacks (or other containers). Supporters of functional literacy programs argue that those who have attended such programs gain more than just these skills. They acquire a self-confidence in their daily transactions that serves them in all areas of life. Detractors of functional literacy, however, argue that the costs of such programs (in teacher time and materials) outweigh benefits, and that most participants immediately forget what they have learned and are not able to read or cipher in any useful way.[1]

Mali is one of the countries which first gave serious attention to establishing a functional literacy program. As early as 1967, Mali established a plan for functional literacy training which was to be carried out as part of programs for 100,000 farmers to increase their production of cotton, rice, and groundnuts. Four thousand "housekeepers" were also to be trained in these programs. Notably, it was assumed that men were the producers, and women were taking care of the home (see Ly 1976).

There were seven district centers for functional literacy programs in Kita (groundnuts), Bamako (industry and cotton), Segou (cotton and rice), Koutiala (cotton), and Mopti (rice). These seven districts were divided into forty-five zones which each have approximately twenty-five to thirty literacy centers. These are run by paid government workers, trained in the techniques of this type of education. The government workers engage *animateurs,* who are literate farmers who teach in their own villages. The villages are responsible for building a literacy center and for its maintenance. In 1973, there were 1,609 centers in the seven areas; 40,777 men and 1,079 women were enrolled (see Ly 1976).

The initial neglect of women in the functional literacy program was corrected at least in part in 1976 when the Women's Promotion Division of the National Institute of Functional Literacy and Applied Linguistics (INAFLA) was established. It is still true that fewer women are trained than men, but in 1983 in three of the seven regions, 3,000 women were participating in the program.

Mme. Dembele Sata Djire is head of this Women's Promotion Division. She is an enthusiastic supporter of the functional literacy program for women and believes that the program has had a significant impact on the lives of the women who have been students.

Note that the functional literacy program for women does not confine itself to training but, as Mme. Djire describes in this paper, frequently has attempted to promote income-generating activities in the villages where it has a training program and has introduced labor-saving technologies, such as grain mills, in the domestic area. It is also interesting that the program relies on modern and sophisticated technical equipment such as video tape machines to transmit its message to the village women.

The Women's Literacy Program in Mali

The Republic of Mali is a landlocked, agro-pastoral country with seven million inhabitants, 51 percent of whom are women. Malian women are present in all socio-economic activities of the country, at all levels, both in rural and in urban communities. However, the great majority of women suffer from illiteracy (around 95 percent are illiterate) and live in ignorance, tied to traditional routine tasks. These conditions prevent them from fulfilling themselves and participating fully in development efforts. Recognizing the danger of this ignorance, the government of Mali set up literacy programs to which women can turn for their education. A

Women's Promotion Division has been set up within the National Functional Literacy Board to undertake research oriented towards the definition of a national policy regarding education and training for women, in order to increase and improve their participation in the development process.

This Division has a drafting section which prepares the didactic material necessary for the women's literacy centers and which writes the specialized booklets used in post literacy training. It also has a training section which trains the Literacy Center's *animatrices* and the female staff of all development services.

At the beginning in 1976, the Women's Promotion Division wrote a program based on the socio-economic activities of women entitled "Women's Participation in Development." It has five large chapters including (1) family care, (2) agriculture and animal husbandry, (3) handicrafts, (4) trade, and (5) wage work. These are divided into two booklets which are used together with a series of pictures during lectures introducing the lessons. Presently these booklets are written in Bambara, Peulh, and Dogon. They will be translated in all the other national languages of the country.

The second chapter of the program, dealing with agriculture and animal husbandry, gives women the necessary knowledge to be informed and trained in these fields in order to produce more and increase their income. The contents of this chapter are: land clearing—showing the necessity of preparatory work; respecting the seasonal agricultural calendar; knowing modern cultivation techniques for maize cultivation and for an early crop which eliminates hunger during the gap period; for millet cultivation which is a basic food crop; for groundnut cultivation which is a cashcrop; and for vegetable gardening and for the introduction of vegetables and fruits in dietary patterns. Finally, poultry and small animal husbandry—to reduce low intake of animal protein and provide some monetary resources—is discussed. This training is reinforced in post-literacy programs by technical brochures concerning agriculture.

Activities of the Division

Literacy programs teach women to read and write through themes pertaining to health, agriculture, handicrafts, education, etc. and to count and do simple arithmetic to solve problems within the framework of their daily tasks. It therefore enables them to better understand the meaning of modern life, to adapt to their changing environment, and increase their productivity. But this educational program faces enormous problems for which the Women's Promotion Division must find solutions. The principal problems are the lack of time because of the burden of women's normal work load (finding water, gathering wood, preparing food, working in fields, pounding millet, etc.); sicknesses of children; men's reluctance to send women to courses; the lack of motivation of women who do not always perceive immediately the rewards of being literate; and the lack of reliable *animatrices* and the fact that the latter are unpaid.

To solve these problems and enable women to have some free time for courses, the Women's Promotion Division has adopted the following strategy:

(1) to undertake programs to lighten women's heavy burden by providing grinding mills to villages, constructing improved stoves, developing dyeing activities, preparing local soap, and creating women's cooperatives; (2) to then show the need for literacy training. It is within this framework that more than 500 women belonging to twenty villages of the Koulikoro region are involved in literacy programs.

These DPF activities are reinforced by the use of video. The audio-visual method has made possible the *animation* (mobilization and awakening) of women in thirty villages of the Koulikoro, Segou, and Mopti regions. Thirty Women's Centers operate in these zones and nearly 3,000 women participate in literacy training activities. This new technique offers women the possibility of knowing others from neighboring villages, of sharing their experiences, and of giving them motivation to find solutions to their problems. The audio-visual programs—based on the women's own needs and including agriculture, health, water and environmental hygiene, nutrition, dyeing, and gardening—are highly appreciated by women who participate directly in making the films and in the debates that follow their showing. This brings about a change in attitude and behavior and leads to important decisions taken at the village level. Video attracts women to participate in programs developed to improve their living conditions and reinforces communication and cooperation between villages. Interesting initiatives taken by individuals or groups which the population can benefit from are made known through video.

But women's training enters into a larger context. Its success depends on a greater coordination of activities of all services in the rural areas, for they all have the same objectives: the development of the rural women's world.

Notes

1. In an evaluation of an Ethiopian literacy campaign, for example, Margarita and Rolf Sjöstrom show that results of such training are variable. Success in terms of learning what is taught depends on the training, and commitment of the teachers (see Sjöstrom 1982).

9 Improving Women's Rural Production Through the Organization of Cooperatives

Sacko Coumbo Diallo

Editor's Note

Cooperatives are encouraged in most Sahelian countries, whatever the political characteristics of their government leaders, although the extent to which they are encouraged and private activity discouraged varies as does the area of activity in which cooperative organization is promoted. The reason cooperatives are popular in Sahelian countries, and even among relatively conservative donor countries who operate in the Sahel (like the United States) is the general poverty of the population. In the rural areas, for example, Sahelian farmers do not have enough money to buy needed tools and inputs. They generally do not have draft animals, much less trucks for marketing or purchasing and transporting supplies. They do not have the collateral to receive individual credit. Nor is there a developed rich farmer *(kulak)* class. The large majority are poor small farmers (living in extended families). There are farm laborers as well who work on government-owned large plantation-type farms or seasonally on the fields of the relatively well-to-do.

In such a setting, cooperatives enable communities to get credit, acquire machinery, obtain needed inputs, sell their produce, etc. They also are a means of starting income-generating activities such as a cloth dying and weaving, and enterprises such as bakery shops in societies where few individuals have the capital to start a business. The government is able to make loans in money or give actual inputs and equipment to the cooperatives which the cooperative can repay much more easily than could an individual who would be ruined for life with the debt if one of the more common disasters—such as drought, ill health, a plague of insects—ruined his crops or slowed down his work after he had received a loan. Operating through cooperatives also can mean a more efficient use of scarce resources and can facilitate communication of information and training.

Cooperatives are particularly appropriate for the Sahel where traditionally individualism was not encouraged. In Sahelian societies, age and gender groups, working together in various domestic and productive activities[1] (such as the *ton*, breeders' and farmers' associations in Mali, mentioned by Mme. Traore) were common.

But, in the Sahel as elsewhere in the developing world, experience with cooperatives has raised some serious questions about how well they serve the interests of their various members. There is evidence to suggest, for example, that in some situations cooperatives actually disadvantage the poorest farmers. Some observers argue, however, that this is not always true and may depend on factors such as the distribution of wealth in the population or the past national history with cooperative organizations (see Ghai *et al*. 1979, 113–158; Lele 1981, 55–72; Esman and Uphoff 1984). In any case, the whole-hearted endorsement of cooperatives as a solution for developing countries by planners of the 1960s and 1970s has given way to a more questioning attitude.

Mme. Sacko Coumbo Diallo does not share any of these doubts. She is the head of the Women's Cooperatives Division within the framework of the Malian national Board of Cooperatives. She is fully convinced of the value of cooperatives for rural women. Whether or not they are thoroughly enthusiastic about cooperatives, Mme. Diallo feels her mission is to persuade women to form such organizations. Her paper is particularly interesting because of this perspective and because she reveals the real problems which are encountered by those who try to encourage cooperatives, such as a basic lack of understanding among rural women of how cooperative ventures should work, poor management skills, and a lack of adequate available resources.

Mme. Diallo's chapter is important because cooperatives have been promoted in Mali over a long period of time, before many of the other Sahelian governments gave much emphasis to them. In the era of Modibo Keita (deposed by coup in 1968. See Section I, pp. 00–00), when the orientation of the government was Marxist, cooperatives were encouraged in all areas. Not only were farmers asked to form producers' and marketing cooperatives, but the government tried to turn the distribution of basic commodities like food and other materials over to cooperatives. Thus UNICOOP was formed in 1964. This was an organization of cooperative stores through which food products would be sold. It is not surprising, given the government's strong encouragement, that 70 percent of the Bamako population bought food from UNICOOP stores in this period (there were 30 UNICOOP stores in Bamako in 1964).

The government of Moussa Traore is much more conservative (see Section I, pp. 00–00) but has not abolished all the cooperative undertakings which were started in the 1960s. UNICOOP, for example, still provided food to 30 percent of the population of Bamako in 1977. But other sources of food were encouraged and private undertakings in others areas invited. Still, it is interesting in this setting where cooperatives have existed over a long period of time to see what they have achieved and what the overall experience has been. Mme. Diallo talks only about women's cooperatives and her Division has only existed for ten years, but the information she provides is nonetheless very useful for those exploring the limits and potential of cooperatives in rural areas.

Introduction

Malian women represent more than 50 percent of the total population and 85 percent of them live in rural communities. The main responsibility of rural women is to take care of their families and their households. They must play the roles of wife, mother, and housekeeper. Furthermore, they fully participate in agricultural and handicraft work. It is essential to find adequate ways to improve their productivity and the lives of this hard-working section of the population.

Women in Production and Trade

In certain areas, women share even the hardest tasks such as ploughing and land clearing. But in general, women sow, weed, watch over the grains close to maturity, winnow, and transport the crop to the house. In addition to their participation in work on the communal field, they take care of their personal plot, growing millet, cotton, groundnuts, rice, *niebe,* etc. They also cultivate small gardens which provide them the leaves and vegetables necessary for the sauces [used to garnish the main meal], and take care of gathering and processing the *karite* nut to make butter, and *nere* to make *soumbala*.

Women usually take care of calves, sheep, and goats. In families with many boys, they are freed from this task and only take care of milking and selling the milk. They are free to make use of the proceeds of the sale of their own cows' milk and of the milk of all the family members' cows.

In fishing, men take care of catching the fish. Processing (smoking, drying), stocking, and selling is left to the women. Handicrafts also occupy an important place in the rural women's activities. They spin, process cotton and wool, make jars, dye, and make objects with straw, leather, wax, and wood. Peulh and Sonrai women are specialized in the making of mats and basket weaving.

Women are very active in trade at the local market level. They are the principal participants in the local market, both as buyer and seller of the products of wild plants, fish, local handicrafts, cooked dishes, cereals, milk products, vegetables, etc.

Problems Encountered by Rural Women

Rural women encounter three major problems relating to land, the means of production, and financing. First, women do not have land which belongs to them; rather they have the usufruct of a field. This right is limited insofar as the family or the community can take away the field when arable land becomes rare, when cash crops increase to the detriment of food crops, or when agricultural projects demand new land distribution.

Secondly, women have few tools. They only use the *daba* (traditional hoe) and do not have access to ploughs, carts, multipurpose motor cultivators, fertilizer, seeds, insecticides, etc.[2] Nor are women trained in modern agricultural methods; even handicrafts are done by hand with no mechanical equipment.

Finally, women cannot find financing, even of the most modest type, for their economic activities.

The National Cooperatives Board

The National Cooperatives Board is conscious of the very important role which rural women play in the socio-economic development of the country and therefore encourages the participation of women in the Malian Cooperatives movement as a way of finding solutions to the problems they encounter. A Women's Promotion Division was created within the National Cooperatives Board in 1975.

The objective of the Women's Promotion Division is to mobilize women to participate in cooperative activities. To this end, activities which are likely to interest women are promoted and help is provided to women who wish to participate. Such help includes finding solutions to financing problems, giving material and moral support, and acting as intermediary with the authorities. The activities sponsored may relate to agriculture, animal husbandry, handicrafts, gathering wild plants, fishing, collecting wood, and savings schemes. The economic activities chosen are tied to the existing cooperative structure in the district.

Within this context, nine women's cooperatives have been created. Four of these are in rural communities and include the vegetable garden cooperative for women of Niono, the hand dyeing cooperative for women in Markala, the women's pottery cooperative of Kalabougou, and the multifunctional cooperative of Kabala.

Urban cooperatives include the hand sewing cooperative for women in the Bamako district, the women's soap cooperative in the Mali district, the agro-pastoral cooperative for Commune VI women, and the women's multifunctional cooperatives for Commune II and Sikasso.

The Women's Promotion Division also supports a program to help provide grinding mills, managed by women's cooperatives, to women in villages.

Objectives and Problems of Women's Division's Work with Cooperatives

The short-term objectives of the women's division include rationally organizing collective work, improving the quality of production to benefit members of collectives . . . , ensuring a regular supply of raw material and production equipment (on the basis of collective ownership), organizing the development of products in conditions of the highest possible quality and best sale price, and increasing and stabilizing the income of cooperatives. In the long run, the Division intends to organize an exchange among towns and villages, develop intercooperative relations between production and consumer cooperatives, improve the quality of production both for the benefit of cooperative members and for the community as a whole, raise the social, economic, and technical level of rural women in particular and of the population as a whole, give each cooperative member the means to defend herself and assert herself within such an association and, finally, transfer planning and management to the women's organizations.

The cooperatives are dynamic institutions which help make women's work profitable. But a number of problems do arise when these institutions are established, and in their management. In the first place, most women are illiterate. They lack the necessary information and education. They lack experience in the field of cooperative management. They do not have sufficient working funds in the cooperatives and, in addition, often do not have necessary basic equipment. The establishment and management of cooperatives also is made more difficult by the individualistic spirit of certain women who prefer to work alone.

The solution to these problems requires the provision of working capital and adequate equipment. It would be helpful if arable land would be made available to agricultural cooperatives and if the work load of women—both domestic and productive—were lightened by the introduction of simple labor-saving technologies. Exchanges (of experience) among cooperatives would be useful. Finally, adequate training is necessary. Functional literacy training within the field of activity of the cooperative would help its members to be able to manage their own enterprises. Specific technical training to better utilize their tools and equipment would also be an important way of solving some of the problems cooperatives encounter.

Examples of Cooperatives

The Women's Promotion Division is convinced that cooperatives provide a major solution to the problems of rural women and offers two examples of working cooperatives already in operation. The first example is the women's multifunctional cooperative of Commune VI which was established in 1975 and was the first of its kind in Mali. It organizes about thirty middle-aged women around agricultural and pastoral production objectives which the women feel will improve their living conditions. However, the results of the work of this cooperative have not been entirely successful for three reasons. In the first place, the women have not adapted to cooperative principles concerning agro-pastoral production. Secondly, grants which were received were badly utilized. Finally, the women lacked means of transportation and had to limit their work to a few farms near

Bamako. For ecological reasons, these farms have poor yields so that the cooperative shows a chronic budgetary deficit.

The Bamako hand sewing cooperative provides a second case. It was established in April 1981 and involves sixteen women whose average age is twenty-two years. All the women are seamstresses. Despite difficulties, this cooperative has had rather satisfying results. On a balance sheet for June to December 1981, there is a gross profit of 943,864 FM with a turnover of 2,488,830 FM. The balance sheet of 1982 shows a gross profit of 485,857 FM with a turnover of 8,376,480 FM. The latter result is rather poor compared to the second half of 1981, but this can be explained by comparatively very heavy expenses (2,210,000 FM as compared to 988,495 FM in the second half of 1981). These expenses included personnel, taxes, income taxes, maintenance, furnishings, exterior services, transport, various management expenses, financial expenses, and depreciation. However, if this cooperative were provided with a workshop and sufficient material for production, had regular and permanent marketing arrangements, and enjoyed better cooperative management, it would undoubtedly meet its objectives.

Conclusion

Rural women fully participate in all socio-economic development activities of their communities. Unfortunately, this contribution is not always taken into account by policy makers when they prepare rural development projects. The analysis of the role of women in production, and of the probable effects of a project on them and their living conditions, is either absent or occupies a very low priority place. It is essential that the economic role of women be recognized and valued, and that development programs take into account the realities of the area and the needs of the population by concentrating on the base level. Cooperative organization must be favored and encouraged in rural communities since it allows rural women to improve their living conditions by producing, gathering, processing, and transporting and marketing their crops, fish, dairy products, handicrafts, etc. in common.

Through cooperatives, women can efficiently defend their interests.

Notes

1. The distinction between domestic and productive is misleading. So-called domestic activities such as caring for children, maintaining the home, gathering firewood, hauling water, and preparing and cooking food for the family are productive activities in the true meaning of the word. What is implied by using this distinction is that productive activities, as opposed to domestic, are to earn cash. Thus, preparing food for sale is productive while cooking for the family is not. The reason for persisting in this terminology is that alternatives are at least equally misleading. It is not quite "domestic" versus "income generating" or "inside" versus "outside the home" that is meant, but a merging of these two.

2. *Editor's Note:* See Dr. Venema's chapter 4 for a slightly different perspective on the access of women to modern equipment and inputs.

10 Appropriate Technologies for Women of the Sahel

Jacqueline Ki-Zerbo

Editor's Note

This paper by Mme. Jacqueline Ki-Zerbo, and the one following it by Mr. Jonathan Tucker, are about the use of appropriate technology by women in the Sahel. The term "appropriate technology" has commonly been misinterpreted to refer only to simple, capital-saving technologies. In reality, most people identified with promoting the use of appropriate technologies mean whatever techniques are most suited to a given situation in terms of local priorities and values, available skills and resources and the nature of the task at hand (see McRobie 1981). Thus, there may be circumstances in which a very sophisticated, capital-intensive piece of equipment or method of production may be appropriate. In other circumstances however, what will be needed is a tool or method of production which is simple, inexpensive, and easily made available to large numbers of people.

Mme. Ki-Zerbo, a trained sociologist working for the United Nations voluntary fund for women, is concerned with the need for technologies of the latter type to lessen the work load and increase the productivity of women in the Sahel. She speaks of a "trousseau of technologies" and her concern is the total life of women, especially in the poorest subsistence communities. Such a "trousseau of technologies" might include the technologies identified by Marilyn Carr as "required to produce the basic necessities of life":

Agricultural production—tools and equipment for ground preparation, planting, weeding, and harvesting, along with basic tools and techniques required for their manufacture (blacksmithing, welding, and woodworking).

Water supply (horticultural)—equipment for storing, lifting, and moving water.

Crop processing—shellers, winnowers, mills, oil extractors, decorticators, fertilizers, and foodstuff manufacture and by-products.

Storage—storage equipment appropriate for different crops using local materials.

Food preservation—metal and glass containers, cooking utensils, equipment for smoking, sun-drying, and packaging of different foods.

Clothing—equipment for ginning, spinning, weaving cotton and wool; manufacture of footwear dyes and finishing materials; tailoring equipment; leather tanning, and manufacture of footwear and animal harnesses.

Shelter—brick and tile making, lime burning, cement substitutes, small-scale cement production, soil stabilization, timber production and byproducts, cast and forged metal fittings.

Consumer goods (not included above)—household utensils; equipment for making pottery, ceramics, furniture, soap, and sugar; cooking stoves; toys; and equipment for water purification.

Community goods and services—school and medical clinic equipment, equipment for road making, bridge building, water supply, power sources and transport (Carr 1978, 9–10).

As Mme. Ki-Zerbo points out, one of the best ways to make new technologies available to Sahelian women is to use (or develop) those which closely relate to traditional technologies but, perhaps, slightly improve them. New technologies, although simple, have been promoted by government agencies after testing outside the locality in which they will be used, indeed even outside the country. This means that there are often problems of adaptation and unforeseen obstacles to widespread adoption of the new way of doing things.

Since poor countries (and certainly subsistence communities) can not afford to carry out the long and expensive process of testing technologies, it is not surprising that these often come to the Sahel from the outside. The crucial step is to test them sufficiently in the local conditions and make the necessary alterations before promoting them strongly in projects. Otherwise people are disappointed and scarce money is diverted into programs to introduce technologies which will never be used outside the project zone.

The improved stove program is an example of the introduction of a technology which would lighten the work load of women with the additional important benefit of decreasing the use of wood in an area in which deforestation is a severe problem. Mme. Ki-Zerbo recognizes some problems in this program with which she was associated in Upper Volta. She is aware that there is not yet a model which can be easily made available to most rural women, but she feels the efforts to teach about currently available models are important not only for the direct benefits which they represent to those who adopt the stove, but also for the corollary impact of the program on developing the role of women in Sahelian society.

Introduction

The title of the contribution that the organizers of this seminar asked me to make was: "Appropriate technologies to help women of the Sahel in their role as agricultural producers." I changed it somewhat, not only to simplify it but for certain other reasons, the principal one being that even though the main interest of CILCA is agricultural production, women of the Sahel are not all, or only, agricultural producers. On top of cultivation as such, fishing, animal husbandry, and handicrafts occupy an important place in the economic activities of sedentary and nomadic women farmers. Furthermore, all women of the Sahel are compelled to do a certain number of daily tasks which affect their economic role and their social condition, whatever the principal economic activity of the community to which they belong.

Since others in this seminar will most certainly speak of technologies and techniques pertaining to agricultural production, in the present paper I would like to emphasize domestic technologies and their possible contribution to a socio-economic change, however small.

Appropriate Technologies: In Search of a Definition

All human societies have developed tools to increase their control of nature and reduce the work necessary for their survival and their subsistence. These tools often are the result of experiences lived and observed with attention and good sense. The whole of knowledge and capabilities developed in such a way constitute *knowledge* and *know-how*. Systematized knowledge expressed through general laws which are valid everywhere regardless of the time factor is called *science*. The application of science to produce goods, services, or ideas leads us to the field of techniques and technology. Modern techniques, mastered by industrialized countries, are often opposed to appropriate technology, which is viewed as a second-class technology, appropriate for backward countries. In truth, technology is not only a way of producing with tools, knowledge and skill. Technology also carries economic, social, cultural and cognitive codes. Therefore *appropriate* technology should be a production process in harmony with the economic, social, and cultural context of products which are appropriate in the same way.

Traditional Technologies

The working day of a woman of the Sahel includes a series of vital tasks for which traditional societies developed more or less appropriate technologies. Water which is so essential to life is rare in certain regions or during certain seasons. In order to procure water, much work is necessary either to draw it or to fetch it from the river or pond, sometimes very far away. The drawing of water is done with a very simple technology: a container is tied to a rope and pulled either by hand or with the help of animals. Water transport is done with locally produced containers (calabash, water jars) usually carried on the head. Calabash or earthenware jars are also used for water storage.

Sahelian food is based on cereals (millet, sorghum, maize, rice) which have to be processed before being eaten by hulling the maize, threshing or pounding millet to separate the grain from the ear or from the chaff, pounding to take out the bran, and pounding or grinding to obtain flour. These various operations are done with a mortar, pestle, or grinding stones.

Sahelian food is, as has already been said, based on cereals cooked in water and accompanied by various sauces which bring additional nutritious elements (proteins, minerals, and vitamins) and especially season the various dishes. Utensils used usually are pottery jars or aluminum pots often made with salvaged tins. Cooking is done on three stones or on pottery or metal stoves which are locally produced. In fact, the Sahelian

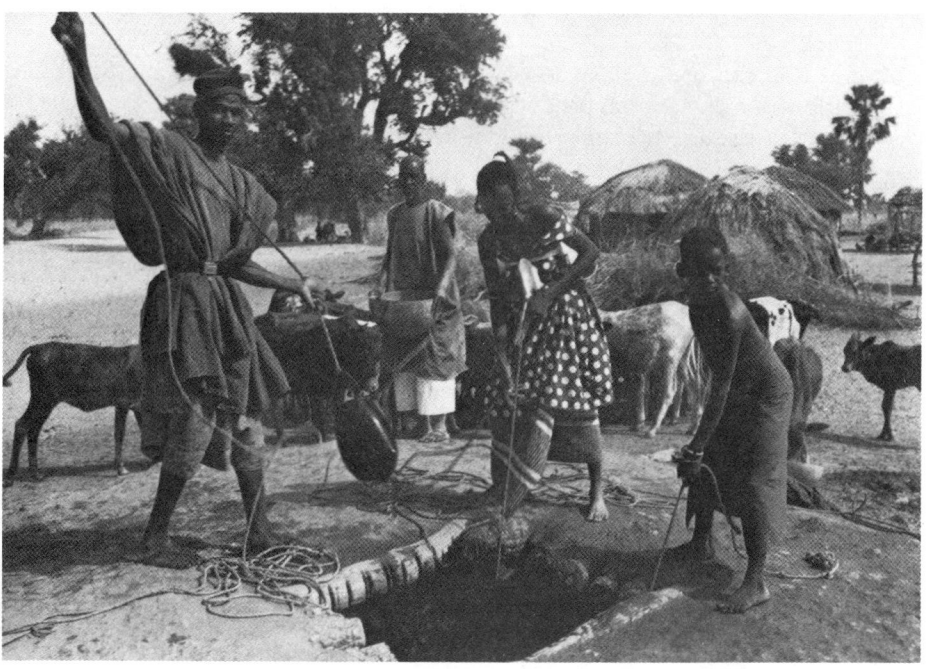

These people are drawing water from a traditional well in Mali. The wells are reinforced with small log platforms and are often very deep. Water is hauled in containers tied to ropes and used for all family purposes, for the livestock and for watering vegetables. Photo by Michel Renaudeau

housekeeper uses a rather limited amount of crockery and utensils: a calabash, enamelled pots which have replaced wooden bowls, ceramic ware, a skimmer, and wooden spatulas to mix the sauce or knead the dough. Lids made of enamel or basket work complete this set of kitchen utensils.

House cleaning consists in sweeping the hut and the courtyard with a simple broom or a reinforced broom decorated with woven ropes. At feast time, it used to be common practice to tamp the earthen floor of the house and cover it with carefully prepared clay to reduce dust and have a cleaner environment. This care for cleanliness and aesthetics was also manifested in the walls which were usually whitewashed and decorated by women. Although there was no particular technology for the washing of clothes, women of the Sahel knew perfectly well how to make soap and starch.

On top of the daily tasks enumerated above, women of the Sahel transform many natural products either for their personal use or for sale. Among these activities one should note: spinning cotton, making pottery, weaving baskets, smoking fish, and making soumbala, karite butter, various drinks, fritters, etc . . . These operations are made using processes and equipment which correspond to local resources.

In regard to food, women are largely responsible for the protection of the family's health. Their knowledge concerning nutrition, medicinal plants, and health care is mostly centered on the mother and child at critical periods of their development such as pregnancy, birth, nursing, teething, and psychomotor development of the child. Apart from wooden dolls, most toys are usually miniature utensils (calabash, jars, rattles) and are made by women.

All these tasks are well known, but it is important to list them for they are the activities which account for the daily survival of family members. Whoever wants to help women must pay particular attention to these tasks because they demand an important part of women's time and energy. It is by analyzing these survival tasks which define the true status of women that one will understand the priority needs of women and their communities as far as appropriate technologies are concerned.

Starting by listing domestic needs and then describing traditional technologies does not mean preaching backward ultra-conservatism. Rather, it serves to highlight an often ignored, often neglected know-how in the rush toward what is new and which is often taken for innovation and progress.

New Needs and Technological Answers

Traditional technologies fit perfectly into the context of rural societies whose social and economic equilibrium they reinforced through respect for the divisions of tasks according to sex and through the utilization of local resources. But they required a lot of time and energy. The opening of the Sahel to the outside world has introduced new needs which demand new methods of work and new technologies to satisfy them. In addition to subsistence agricultural production, it is necessary to cultivate cash crops such as cotton, groundnuts, tobacco, and vegetables to obtain ready money. Medical care can no longer by provided only by healers, and modern health care requires time to go to the dispensary and money to buy medicine. Education, which was totally free in the traditional system, now occupies an important place in the family's budget because of clothes, school supplies, and transport of school children. Finally, the rural exodus to town, the trend toward urbanization, and the degradation of the environment all tend to contribute to the disappearance of certain products and certain technologies which must urgently be replaced.

All these changes influence the work conditions of women and certain new technologies have already been tried to answer the new needs. For example, seed drills, ploughs, carts, millet grinding mills, hand and solar pumps, improved stoves, and biogas installations are among the things which can teach us certain lessons for the future of appropriate technologies for the Sahel. Often these technologies have been introduced as finished products which the population had simply to use. This results in a certain distance between the technologies and the local environment whose resources do not enter into their making. Furthermore, training is required to make them work; this training is given only to a few members of the community, thus making the others dependent upon them. Upkeep and repairs of the new machines bring about problems which are difficult to solve, so that many technologies introduced in villages are symbols of dreams and hopes soon destroyed.

Lessons and Future Perspectives for the Development of Appropriate Technologies

Technology, as we have said, should be a way of producing appropriate products using local resources, both natural and human. More than the final product, it is the process of making it and its introduction that count. It is important first of all to identify needs and seek solutions in the family and communal context in order to find out precisely how these needs have been satisfied until now. What were the tools and materials used to satisfy these needs? Who used them and how did they procure the material they needed? It is then necessary to study the modifications which could be brought to existing technologies before substituting completely new technologies which are totally foreign to the area. This step may seem slow, but it guarantees attachments and a strong base for the new undertakings. It allows for the participation of users and facilitates the acceptance of new technologies.

Although it is not always possible to promote the local manufacture of many technologies, one can at least encourage a dialogue between inventors, manufacturers, and users so that none of these groups reduce others to the simple condition of passive consumers. It is therefore important to start with pilot operations on a small scale where research and development go hand in hand and mutually enrich each other. Pilot operations must include the training of craftsmen involved in the making and the upkeep of the appropriate technology, and supervision and follow-up activities to find out difficulties in use and upkeep in the field. It is only through these supervision activities that the effective use of the technology, and its real efficiency in terms of reduction of time and energy, will be verified. These supervisory activities also will permit the input from users and innovations brought by some craftsmen, thereby preserving personal initiative and encouraging popular creativity and participation.

Pilot operations are necessary to find out the potentialities and the obstacles which can favor or hinder the development and popularization of a given technology. Through them, it is possible to program the necessary training and to make choices concerning manufacturing systems, marketing, and after-sale services.

The scale and complexity of problems brought about by the introduction of an appropriate technology is such that this introduction must be carefully prepared whatever the urgency of the problem to be solved. The adoption of an appropriate technology cannot depend on the desire outside the community that one wants to help. That is why the *animation* and mobilization of policy makers and local people are at *least* as important as the technical perfection and the production of appropriate technologies.

Training of Rural Women Extension Workers for Appropriate Technologies: Improved Stoves

These statements lead us to consider the role of women extension workers in research for appropriate technologies in the communities in which they are called upon to operate.

The experience of the past 20 years has proven that rural development will take place in an integrated framework or will not occur at all. Isolated agricultural extension, health education, and literacy projects have shown their limits. The approach to the rural world must be totally modified along two main avenues. (1) The *animatrices* must work with the population on all the problems and needs. These can be isolated for the purpose of analysis, but are in reality totally tied to each other. (2) The *animatrices* must collaborate with other development agents to solve the problems of the communities in which they live.

In the case of the improved stove, it is clear that easing the task of gathering wood is a strongly felt need in rural communities because of population increases, drought, and deforestation. The need to reduce the time and energy spent by women for gathering wood is inseparable from the problem of the degradation of the environment which affects the whole of the population. Therefore, the supply of fuel wood and technologies to reduce the amount of wood needed concerns not only women, but the whole of the village community, and even the nation whose help is necessary.

Solutions should be sought and put into effect in the village community. This should be done with the help of political leaders and technicians within the framework of reforestation activities at the village level, construction of improved stoves programs, and functional literacy and health education. Improved stoves are a seemingly simple technology which, however, require an enormous amount of research to conceive and develop models which are resistant, durable, efficient, cheap, and adapted to the culinary habits of housekeepers.

The pilot operation of the stove project, started in 1981 [by CILSS], has not yet led to the selection of types of stoves ready to be mass produced and marketed on a large scale. Nonetheless, some speak casually of more than 100 million improved stoves to be constructed in the next 20 years in developing countries. Although there is a huge market for improved stoves, we must limit ourselves to specific development tasks which we wish to undertake with rural women, step by step. The training workshops for the construction and use of improved stoves have until now been opportunities for the gathering of a very large number of people concerned with the problems of protection and improvement of the environment, energy issues, and problems pertaining to women. Those who participated were mostly forestry experts, research workers in new energies and renewable energies, staff members of the women's promotion technical services, and representatives of youth and women's organizations. The seminars were not only occasions for discussion of deforestation and heat transfer problems, but they also gave women and men the

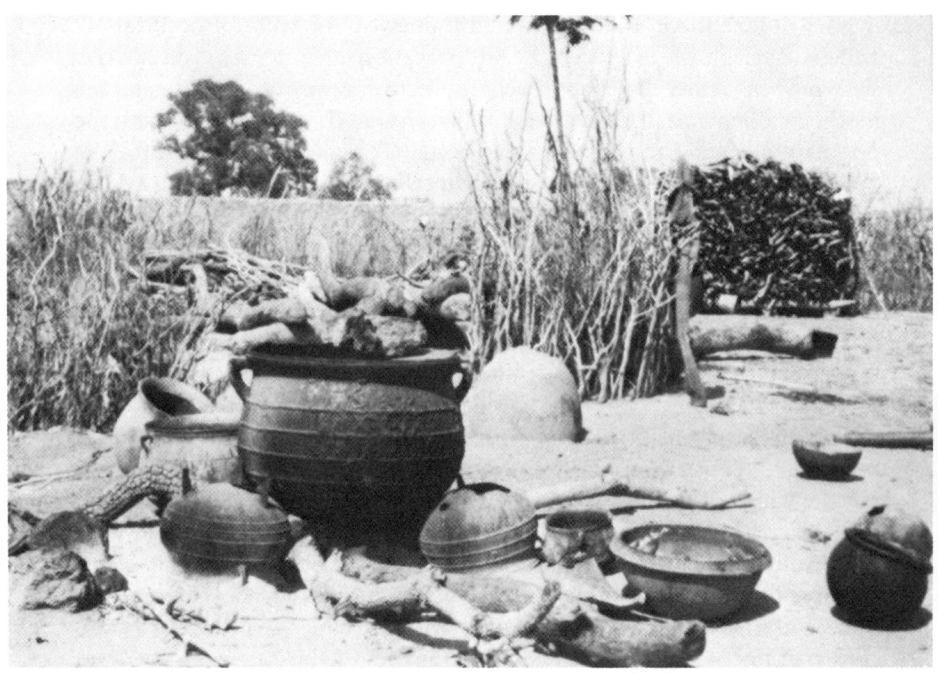

This is the type of pot used over the traditional open fires, which are still more common than any improved stove in the rural Sahel. Sahelian food is based on cereals cooked in water and accompanied by various sauces. Utensils usually are pottery jars or aluminum pots often made with salvaged tins. Photo by Bernhard Venema

opportunity to actually pound and sieve together the clay and sand, to transport together water and bricks, to build together improved stoves, to proceed together to the finishing of the product, and to give together cooking demonstrations.

Without having too many illusions about the eventual changes that the improved stove seminars could introduce in the behavior of male participants concerning the kitchen environment and the preparation of food in their own family, it must be noted that improved stoves are a new field of collaboration between men and women of the Sahel. They are therefore a way to introduce much more profound changes than those which simple thermal performance and fuel economy can bring.

The seminars also have brought about the active participation of women leaders of the rural world who, from Mauritania to Chad, have revealed knowledge of the situation and their capacity to participate in solving their problems.

Many *animatrices* who participated in improved stoves sessions have learned to construct improved stoves, but very few have become involved in making them available to the mass population. However, we know of cases in Mali, Niger, Upper Volta, and Chad in which the knowledge acquired in terms of improved stoves has effectively become part of the program of the *animatrices*. The most encouraging results have been noted among rural craftswomen who, once trained, have constructed improved stoves for themselves and their friends and have trained other women. The most typical cases are found in the Louga region of Senegal and in the Po region of Upper Volta.

The development of portable ceramic stoves opens a new avenue for women who traditionally have been involved in pottery making. Experiments with these stoves undertaken by UNFM (Union Nationale des Femmes du Mali), with Canadian help, merit close study.

Improved stoves offer a favorable area for several rural development programs such as nutrition education (with the demonstration of improved mash cooked on improved stoves), sanitation education (with the presentation of improved stoves as being part of a "trousseau" of appropriate technologies including latrines and water filters), and the functional literacy programs (with improved stoves as a center of interest for the preparation of didactic material and practice sessions).

Finally, it is the choice of construction material, type of stove, and system of production which will determine the degree of women's participation in setting up and popularizing improved stoves.

To satisfy the huge needs felt in the Sahel, the development of varied and diverse technologies is necessary. However, it is important to consider—over and above the tools, equipment, and products offered to consumers—the contribution that the technologies can bring to the start of a development process sustained and enriched by the initiative, creativity, and march towards progress of individuals and communities.

Such appropriate and participative technologies are, without a doubt, more difficult to conceive, set up, and popularize than others. But lessons of the past show us that such is the price of development.

11 The Lorena Cookstove: Solution to the Firewood Crisis?[1]

Jonathan B. Tucker

Editor's Note

This paper is the only article in this book which does not directly deal with women in rural areas. It is included here to provide a second perspective on appropriate technology and its relevance to women in the Sahel.

Jonathan Tucker, a doctoral candidate in political science, is a freelance writer living in Boston who specializes in science, technology, and Third World issues. He has recently returned from working for CARE in Somalia.

Mr. Tucker was never directly involved in an improved stove program in the Sahel. He has, however, reviewed the many and lengthy reports, articles, and documents on the introduction of improved stoves, specifically the Lorena model, in the Sahel and elsewhere in the world.

Mr. Tucker's contribution to this book is a scholarly overview of the positive and negative factors associated with the dissemination of the Lorena cookstove and its derivatives. His conclusions are more negative than those of Mme. Ki-Zerbo. "Appropriate" (simple and labor intensive) technologies are needed in the Sahel but, according to him, it's very difficult to develop them correctly for different local situations or disseminate them broadly.

Looked at closely, the facts cited by Mr. Tucker are not different from those presented by Mme. Ki-Zerbo. Both observe the same basic situation and both identify similar problems. Mme. Ki-Zerbo, however, stresses the positive aspects of appropriate technologies for rural women, while Mr. Tucker is more concerned with demonstrating the difficulties in their adaptation and dissemination.

Introduction

The growing demand for energy in the form of wood in many developing countries poses a major threat to the fragile environment upon which these nations must depend for their survival. In the arid Sahel region of Africa, just south of the Sahara, firewood currently provides 94 percent of rural energy needs. Because the consumption of firewood consistently exceeds supply, tree depletion is taking place at about twice the natural rate of new growth, causing erosion of the topsoil and progressive desertification.

It is common for women and children there to travel more than thirty miles to collect firewood. Some poor villagers have been forced to do without fuel, eating raw millet flour mixed with water instead of preparing the traditional pancake-like bread. Because the dough is not cooked, fewer nutrients can be absorbed. The villagers are also unable to boil their water and hence have become more vulnerable to debilitating waterborne diseases and parasites.

Unfortunately, there are no readily available substitutes for wood as an energy source for the rural sector. Although some villagers have switched from wood to cowdung cakes as cooking fuel, the use of dung is not economical, given its low heat efficiency and the time and labor required to collect it. Burning dung is also ecologically unsound in that it reduces the fertility of the soil.

Other more exotic technologies, such as solar cookers and biogas plants, are currently too expensive for family use and are more likely to be employed in small-scale industrial processing rather than in individual homes (Ki-Zerbo and de Lepeliere 1979, p. 8). It therefore seems likely that wood and other biomass fuels (such as millet stalks) will remain the primary source of energy for the rural poor in the Sahel for some time to come.

In order to bring wood production and consumption back into balance, there is an urgent need for reforestation programs and the establishment of village woodlots. Such programs will take years to come to fruition, however, so the only short-term hope for an amelioration of the crisis is to reduce the end-usage of wood. In Upper Volta, present fuel consumption is at the exorbitant rate of one kilogram of wood per individual per day, largely because of inefficient cooking methods (Vita 1980, 2).

The traditional hearth in the Sahel consists of a triangular arrangement of three stones over an open fire. Due to the unprotected flame, only 5–10 percent of the potential energy contained in the firewood is used, with the remainder being lost to the surrounding air. Introducing efficient wood-burning stoves for domestic use might therefore have a significant impact on the rural energy crisis, provided that implementation is rapid and nearly universal in scope. Even a 10–20 percent reduction in the consumption of firewood could save millions of trees and slow the pace of deforestation significantly, providing time for natural growth to catch up with the cutting rate.

There are a number of other rationales for the introduction of efficient cookstoves, particularly the health hazards associated with cooking over an open fire. Because the fire is at ground level, the cook or her young children may be repeatedly burned or scalded, and the smoke can cause chronic eye and lung irritation, leading to blindness, bronchitis, emphysema, and lung cancer. Flying sparks create a constant fire hazard, and the smoke blackens the cooking area and much of the house with soot.

Smokeless cookstoves could therefore have a major beneficial impact on public health. They would also reduce the time and effort devoted to gathering firewood, freeing up resources for investment in development-related activities. For example, women could allocate the time saved searching for wood to cottage industries that would increase the family's income.

The Lorena Cookstove

The design for an efficient woodburning cookstove originated in India in the early 1950s and was based on the traditional Indian stove, or chula. The Hyderabad chula, developed by the Hyderabad Engineering Research Laboratory, provided the inspiration for the more advanced Lorena stove, which was widely promoted in Guatemala, Honduras, and other Central American countries during the 1970s.

The Lorena stove is sophisticated in concept but easy and cheap to build, requiring only hand tools and locally available materials. It works by containing the heat of the fire and channelling it through a network of internal flues that concentrates the heat at the cooking holes.

The basic construction material is a mixture of sand and clay called *lorena*, a hybrid of the Spanish words for clay *(lodo)* and sand *(arena)*. Because the Lorena stove is massive, it must be built on the spot where it will be used. Hence, there can be no central production or distribution of the stoves; each one must be made by hand by the owner or by local craftsmen.

To build the stove, a large block of lorena is first formed and shaped with hand tools while it is still wet. A horizontal tunnel is then carved through the block, and holes to fit cooking vessels are cut through the top. A firebox is constructed at one end of the tunnel, and a chimney installed at the opposite end to carry away the smoke and provide a draft. As the wood burns in the firebox, the hot gases and flames flow through the tunnel and under the pots before being drawn up the chimney.

With a sheet metal chimney and dampers, the Lorena stove costs between $10 and $15 when owner-built and about $30 when built to order. The only major expense is the sheet metal chimney, and this could be reduced by finding cheaper substitutes. The stove itself can be made large or small, high or low, and can be modified to burn a variety of fuels, including wood chips, sawdust, and grain stalks (Banerjee 1981, p. 5). According to Ken Darrow of Volunteers in Asia, "The Lorena stove is more a concept and a material than a particular stove design. Thus, widely varying cultural adaptations are possible" (Darrow 1979, p. 5).

Obstacles to Acceptance

Although cookstoves have been developed that work extremely efficiently in the laboratory, the real test of an "appropriate" technology is for it to be transferred successfully to the village context and adopted by the local people as an integral part of their daily lives. In spite of the obvious utility of the Lorena stove, it has suffered numerous failures of acceptance. A recent study by the U.S. National Academy of Sciences came to the disheartening conclusion that in no area where efficient cookstoves were introduced has there been a significant, long-term improvement in fuel savings. In many cases the stoves have been adopted initially for as long as a year or two, only to fall subsequently into disuse.

Why has this seemingly sensible effort failed so far to achieve its objectives? The biggest obstacle to the dissemination of efficient cookstoves is the financial investment required on the part of villagers. Although peasants are usually willing to invest in new technologies if the perceived payoff is good and the risks are low—for instance, in buying a bicycle—the benefits of efficient cookstoves are not as obvious.

Moreover, women are the primary beneficiaries of the technology (since it is they who must cook and collect firewood), yet it is the men who have the final say in the decision to buy. As women have very low incomes from small businesses and crafts and are therefore unable to pay for the stoves themselves, they require the financial assistance of the husband or a male relative. Women's time is not highly valued, however, and it is only when the search for wood becomes so time-consuming that they are kept from performing other domestic tasks that men perceive wood-conserving cookstoves as an attractive investment.

For families that live in towns and must buy wood, the economic rationale is more compelling. Anthropologist Jacqueline Ki-Zerbo has estimated that a family purchasing 3,000 francs worth of wood each month will recoup their 5,000 franc ($14) investment in an efficient cookstove within four to five months. Still, the amount of wood saved is not immediately obvious and only becomes apparent over a period of time. Thus, there must be motives other than purely economic ones—such as aesthetic, health, or status considerations—to justify adoption of the stove.

In addition, there may be problems related to construction and maintenance. Cookstove design is fairly critical; so that if the firebox is built too high, the stove may burn more rather than less wood. The users may also fail to clean out the flue, with the result that it gets clogged with soot or pieces of food, reducing the draft and causing smoke to leak out the front of the stove. Finally, the villagers may become disenchanted when the appearance of the stove worsens with continued use, particularly if aesthetics or status considerations were primary motives for adoption. Small cracks often appear in the surface of the stove after several months. Although they have no effect on function, the users may conclude that the stove has broken and stop using it.

Cultural factors have created additional obstacles to the adoption of cookstove technology. As Nicholas Jequier has pointed out, introducing new cooking methods is a difficult task because it impinges on "spheres of human activity that are inherently very stable and that tend to be greatly influenced by tradition, ethical norms, religion, and taboos." Western stove designers often overlook the fact that the traditional fireplace, in

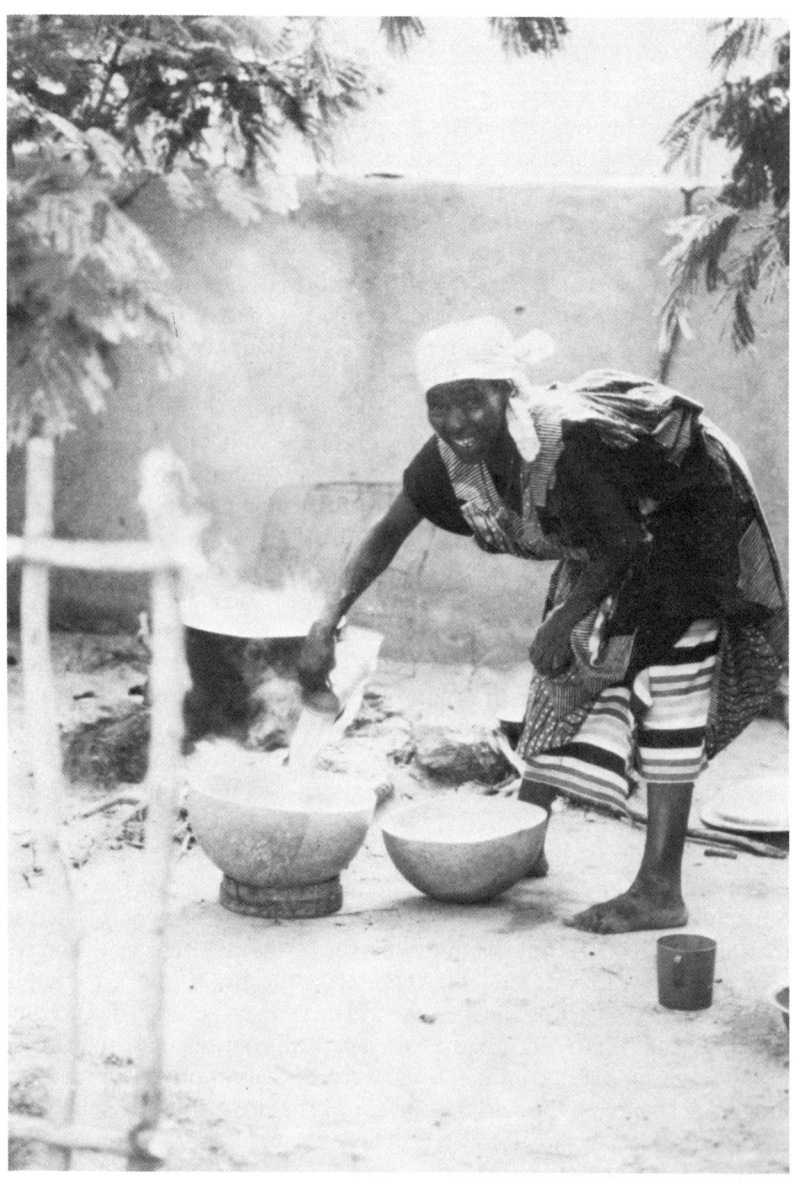

This Bambara woman in Mali is preparing food for a large visiting group in the village of Toko near Segou. The traditional hearth consists of a triangular arrangement of three stones over an open fire. Due to the unprotected flame, only 5 to 10 percent of the potential energy contained in the firewood is used, with the remainder being lost to the surrounding air. Photo by Richard Harley addition to providing heat for cooking, serves a complex of functions within its setting. If these extended or latent functions are not met by the cookstove or provided by alternate means, the technology is not likely to be accepted permanently.

What are the benefits of an open hearth that are foregone by the adoption of an efficient cookstove? A few examples follow.

1. The first drawback of cookstoves is their lack of flexibility. Unlike the open fire, the cookstove has holes of fixed size that limit the number and variety of pots. Although the reduced fuel requirement of cookstoves saves time and effort expended in the search for firewood, the wood must be cut into much smaller pieces in order to fit them into the firebox.

Moreover, the three-stone hearth is portable, whereas the massive Lorena stove is fixed in one place inside the house and cannot be moved. This drawback is particularly acute in areas such as northern India, where cooking outdoors is favored during certain times of year. As a result, although Lorena stoves have been widely adopted in southern India, where women take pride in their kitchens and do all their cooking indoors, the stoves have not been successful in the north (Banerjee, 1981, p. 8).

2. Many villagers believe that food cooked over an open fire tastes better than that cooked on a smokeless stove. Whether the difference is real or imagined is not clear, but taste is more important than many outsiders believe. In Upper Volta, for example, villages attempt to boil water in clay pots—a nearly impossible task—because they do not like the taste of water boiled in more efficient aluminum pots. As a result, they often fail to bring the water to a full boil, and hence are more likely to develop waterborne diseases.

3. Although cooking smoke has adverse effects on health, it does serve a number of useful functions. In many countries villagers depend on smoky fires to keep their thatch roofs dry and insect-free, to drive away mosquitoes, to protect the beams of the house from termites, or to preserve ears of corn and other stored foods from destruction by pests. If smokeless cookstoves are to be adopted in such areas, it may be necessary to introduce, at the same time, low-cost methods for the preservation of wood and thatch, mosquito netting, etc.

4. Unlike the Lorena stove, which is designed expressly to contain the heat of the fire rather than radiate it, the open hearth emits both heat and light. In temperate zones, the fire may be valued as much for space heating as for cooking, and in villages without electricity or kerosene lamps, it may be the only source of light in the home at night.

The traditional hearth may also served as a festive center for social interactions, such as gatherings of the extended family. Where Lorena stoves have been adopted, these latent functions of the cooking fire have frequently led to "utilization flaws." In modifying the technology to more closely meet their needs, the villagers may render it less efficient, thereby negating its major intended benefits, namely fuel conservation and smoke elimination.

Peasants in Guatemala cook directly over the flame in the firebox of the stove rather than using the pot holes for the slow, simultaneous cooking of different dishes. Instead, the pot holes are merely employed to keep cooked food warm—a highly inefficient use of the technology (Shaller 1979, 84). In Botswana, where the traditional hearth serves as a social center, villagers open the front of the firebox to provide heat and light, and therefore do not conserve wood.

In sum, the latent functions of the traditional cooking fire must be taken into account if more efficient cooking technology is to be successfully disseminated. The current design of the Lorena stove should therefore be modified to approximate the same complex of benefits offered by the open hearth: heat, light, and community, as well as a cooking flame.

Factors Which May Encourage Acceptance

Additionally, the basic cookstove design should have enough inherent flexibility to allow for modifications by local craftsmen without significantly impairing its operational efficiency. Guatemalan Lorena stoves, examined after a year of use, were found to incorporate a number of ingenious modifications, sometimes added by the builder but more often by the user. These improvements included an external damper system, a sloping flue, a surface coating of clay-cement mix for improved appearance and durability (Aprovecho 1981, 4–5). Such modifications should be tested and publicized so that they can become widely available.

Because of variation in local needs and cooking practices, the design of fuel-efficient cookstoves must be adapted to the specific circumstances prevailing in each region. In this process, user participation in the design and installation of the stoves gives much better results than a more "top-down" approach. For example, the height of the stove can be adjusted to meet local needs and customs. Women in some cultures like to squat while cooking, whereas others prefer to stand.

A key factor in the success or failure of a new technology is the way it is promoted. Because the population of the Sahel is diverse, a range of stove models and promotion techniques will be required to achieve maximum possible diffusion.

The promoters will also have to take political factors into account, such as the need to secure the approval of local elites and the possible existence of interest groups that oppose the innovation. One source of opposition to the fuel-efficient stoves may come from local entrepreneurs who sell wood at high prices and who may profit from the fuel shortage (Gould and Joseph n.a., 11).

It will also be necessary to train large numbers of extension agents to bring the technology to the people. According to Ki-Zerbo, villagers may entertain unrealistic fears about cookstoves, such as the belief that the stoves will erupt in flames if incorrectly used. Extension agents should therefore be available to demonstrate the stoves, educate the villagers about their workings, and provide clear instructions concerning installation, use, and maintenance.

Because there is a shortage of skilled manpower in Africa, additional extension agents will have to be trained if cookstove extension is to become self-sustaining after the handful of foreign consultants leave. A major failing of current training programs is that, for cultural reasons, women are rarely involved in extension work and almost never participate in the actual construction of Lorena stoves in the home. Since acceptance of the technology may depend on the extent to which the users (all of them women) feel involved in the decision-making process, it is essential to train more female extension agents.

A second problem is that cookstoves are an "invisible" technology, hidden away in the privacy of a family's kitchen, making it difficult to promote the stoves through demonstration. One solution would be to provide model cookstoves for the preparation of fancy foods in the village markets. Finally, a near-complete diffusion of cookstoves throughout the rural population will be required if the technology is to have a significant impact on the fuelwood crisis.

In Upper Volta alone, the German Forestry Mission has estimated that to achieve a sustainable rate of firewood consumption, the country would have to build 250,000 cookstoves per year, or a total of 1 million new stoves over four years. This daunting figure suggests that it is unwise to rely on efficient cookstoves alone as a solution to the fuelwood crisis. Instead, cookstove extension should be integrated with other intensive programs designed to increase wood supply, reduce consumption, and meet basic human needs.

One way to reduce fuelwood consumption would be to establish food-processing industries at the village level that would render food grains softer and easier to cook. Solar stoves might be used to steam newly harvested rice or millet, removing the oils so that the grains last longer and are easier to hull. It might even be possible to process rice and millet into a form of "minute rice" or "instant hot cereal" that does not require much fuel to prepare.[2]

Experience with the Lorena stove has shown that there are major constraints on the dissemination of "appropriate" technology in developing countries. Proponents of small-scale technology, having experienced numerous failures of acceptance, are now moving from a total preoccupation with technical hardware to a new concern with the "software" necessary for the diffusion of technical innovation. In addition to managerial skills and promotion techniques, successful extension requires a profound understanding of a particular society's economic, political, and cultural characteristics, to which the technology in question must be adapted. Unfortunately, what is technically possible will be limited by what is culturally acceptable.

Notes

1. This chapter was originally published in *Environment* 25, (April 1983, pp. 3–5).
2. Interview with Marilyn Hoskins, Department of Anthropology, Virginia Polytechnic Institute, Blacksburg, Virginia.

Conclusion

Topics for Debate and Reflection

Most of the papers in this book use the same development language and make many common assumptions about what is needed in programs for women. All contributors to the book believe that rural women in the Sahel perform essential roles in agriculture and that they have been neglected by policy makers concerned with agricultural development. All contributors place a consistent emphasis on the importance in women's programs of mobilization, *animation,* and training. Most papers stress the need to convince authorities at all levels to back their programs and to persuade male leaders as well as women's groups at the village level to support them. Villages should choose for themselves—so everyone agrees—which women will be specially trained to come back and lead village efforts. All indicate that the success of any program depends completely on whether local women fully participate in the program and are sufficiently trained to carry out the responsibilities which the new programs entail. Most contributors also agree that the whole process of development is so complex that only a multi-faceted approach, taking into account the diversity of functions performed by women, will be effective. Thus, development programs have to consider health, nutrition, and domestic chores as well as income producing activities and regular agricultural tasks.

Illustrated in this book, however, are as many areas of disagreement—implied or open—as there are points of agreement. In the first section, as discussed in the Commentary, debate centered on the extent to which women are automatically disadvantaged in the process of modernization. Dr. Cloud and Mme. Thiam stress the undermining of rural women's economic and social position as development occurs and especially as a result of government programs. Professors Venema and Creevey, on the other hand, argue that it is most useful not to presume a blanket result from each program in each place. Rather, policy makers must carefully review the factors in each situation. Which women—according to such variables as ethnic group, location, age, wealth, marital status, etc.—will benefit or lose, and what will be the impact on the entire family system of any proposed change?

In the second section of this book, three major issues emerge: (1) which area of women's activities should be addressed first? (2) how many areas should a project undertake? and (3) to what extent should projects try to change the lives and customs of project participants?

The question of which activity to begin with in development projects has been debated within most planning agencies in the Sahel. Current outside criticism is that when assistance was available for women, it was for traditional female tasks like child care, family health and nutrition, and domestic chores, rather than for the economic activities in which women engaged. The economic role of women was undervalued, according to this view. But Jacqueline Ki-Zerbo has a different perspective. In her verbal presentation at the 1983 Bamako Workshop, she warned that assisting women in increasing their productivity in agricultural activities should not be done without considering their domestic functions. Lightening their burden in the latter area is a precondition for increasing their productivity in agriculture or other activities. Otherwise they will not have the time or energy to pursue new programs. Furthermore, she feels that domestic technology programs can have a siginficant impact beyond the limited area of that particular technological innovation. Thus, she argues that the improved stove program went beyond meeting a clearly felt need to reduce wood gathering and cooking time. The collaboration between men and women which took place in the program, according to Mme. Ki-Zerbo, and the additional independence gained by women from making the stoves themselves and using them, helped in the process of social development and improvement of relationships between the sexes. Further, the stoves could be part of an improved nutrition program and thus have a function beyond their limited cooking purpose.

Mme. Djire does not disagree that domestic improvements are necessary—in fact her program introduced grinding mills in the villages where it was active. But she passionately defends the seminal importance of functional literacy programs as a primary development tool—a means of teaching other activities and mobilizing rural women to be responsible for themselves in all areas. In Mme. Traore's project, in contrast, functional literacy was not equally emphasized, and in discussion she pointed out that this was because functional literacy was not perceived by rural women as equally important to other activities.

Mme. Diallo writes of a program which gives assistance to women in their productive activities only if they form a cooperative and, indeed, sees her task to prepare a "cooperative mentality" even when women are not accustomed to working cooperatively and prefer individual endeavors. Thus, cooperative formation is given precedence over productive activities *per se* (and by implication over functional literacy and/or domestic technologies).

In essence, no one is arguing that the activities of other projects are not useful, just that their own area is of fundamental importance for development programs. There would be no debate if resources and manpower were unlimited, for then everything could be done. But, as these factors are severely limited, choices have to be made. Should programs stress functional literacy over agricultural production? Should all programs begin with innovations in the area of domestic technology, or should this be secondary to working on improving production techniques or literacy education? Are cooperatives a necessary organizational structure which must be set up before improved production or new income producing activities are contemplated? This book provides no answer. Nor is it adequate to simply ask local women what they want. They want assistance in all areas of their life and some activities may well have broader impact than others and may be needed as a base for further development.

The solution may be to move in steps—begin in whichever area the planners (in consultation with local women) wish to emphasize but be closely aware of other areas of concern and move into these as local need, demand, and project experience indicate. This, of course, begs the question of which is the most fundamental area of intervention, and it leads to the second problem area indicated in these papers: the scope of activity of any one project. Most of these papers illustrate what might be called an "integrated" approach to rural development for women. None of them closely restricts itself to one function performed by rural women. Mme. Traore does argue for specialized training for village extension workers as they are semi-educated women and cannot be expected to learn everything at once. But her project, nonetheless, is training in many subjects and expects to send back many different sets of extension workers specializing in different subjects to the same village area. Mme. Kantara's paper, however, suggests what happens when you try to do everything in the same project simultaneously. Her project has a catch-all approach including everything from introducing domestic labor-saving technologies, to providing equipment and training for agriculture, and credit and marketing schemes. Various objectives of her project appear not to have been met, in part because there were not enough *animatrices* available to disseminate the new ideas and work with local women to bring about their adoption. Motor driven grain mills, for example, were to be spread but could not be until mechanics (skilled in their use and repair) were trained. It is also easy to see from this project the difficulties of project evaluation when so many activities are begun together. Leaving aside possible donor interest in *measurable* progress, there is also a danger that the local women participants will lose sight of what they are actually gaining and be distracted by elements which do not succeed or cannot be achieved as quickly as proposed at the outset.

How many activities, then, are appropriate for one program or project? Certainly no more than there is trained staff to manage. However, it is a matter of judgment as to how many that implies.

The third major topic for debate coming out of these papers (and the discussions which followed them at the 1983 Bamako Workshop) is the extent of deliberate intervention in the lives and customs of project participants. All the Sahelian contributors exhibited their respect for customs and traditions in their countries, and their belief that no project would be effective that did not take into account local values and ways of doing things. Thus, for example, all project staff visited local traditional authorities. But, carrying out a development project implies changing things and that this is desirable. If the lives of rural women are to be improved and their productivity increased, a transformation is expected. The introduction of new tools and techniques, of labor saving devices and literacy programs, will change the parameters of rural women's lives. Given this fact, how much alteration is desirable? Programs or projects can adhere as closely as possible to existing customs while pursuing specific project objectives or deliberately try to use their project to bring about broader social changes. To do the latter, of course, is a "top down," more or less paternalistic, approach. It says essentially: "We know how you should live and we will try to make you live that way." This choice seems undesirable but, realistically, all outside intervention tends to be "top down" by the very introduction into the project area of expertise and resources controlled by individuals from the educated and/or trained elite.

Critics of Mme. Diallo raised the question of whether cooperatives should be promoted so strongly if they did not in fact grow out of the traditional work patterns of the local area, whether or not such an effort to alter work habits might detract from the success of the productive activities envisaged. Mme. Diallo's own presentation, in fact, indicated problems with certain projects because the women participants did not "adapt" to "cooperative principles."

Unintended Impacts of Development Projects

There is no answer in the papers included here to the latter question or, indeed, to either of the other two debates discussed above. The costs and the benefits of the decisions in each case have to be carefully weighed. Perhaps the only thing to be said here is that a conscious awareness of alternative strategies is needed to prevent blind imitation of other projects without evaluating the results these have achieved. There *are* choices. All too often planners do not seem to know this.

Sometimes, however, even if planners consciously try to make a careful choice of strategy adapted to the particular time and circumstance, their projects will not have the impact intended once they are carried out. This is true because factors outside the control of the project staff (such as a major drought) intervene. Sometimes the problem is simply that the real life situation is so difficult that any choice is made at high risk. An example of this is the question of whether a project should give supplies and equipment, or should loan them to project participants with eventual repayment expected. Some planners contend that giving rural women supplies makes them dependent and does not develop initiative and responsibility. But others state that rural Sahelians are so poor and so vulnerable to such things as drought, raiding animals, and insect pests that loans (even small ones for basic equipment) pose an intolerable burden. What happens to the women in Mme. Kantara's project, for example, whose donkey, received on loan, is eaten by hyenas?

In the latter case, the planners did not intend to saddle already poor women with a burden, but the result of their project was that the women have a debt they cannot pay. Even more serious is the situation in which planners draw local participants into projects or programs when information on the real problem or the actual consequence of the improvement is inadequate. One example of this in these papers is the case in which technical information is not complete on the impact of the introduction of laborsaving technologies. Although Mme. Ki-Zerbo writes with great enthusiasm of the "trousseau" of appropriate technologies needed for women in the Sahel, these sometimes have not worked out as was intended by planners. Thus Mme. Kantara talks about motorized grain mills in her project which cost more than an individual rural woman, or a group of village women, could be expected to pay and require a machanic for their maintenance or repair. In her verbal explanation she said that hand mills had been tried first, however, rather than being laborsaving, they were difficult to use and very tiring. Perhaps this particular technological advance had not been developed sufficiently to be appropriate to local needs, at least at the time of her project.

Jonathan Tucker suggests that the stove programs also have not been completely successful in the Sahel (although some more so than others). The Lorena stove in particular turns out to have all kinds of problems in construction, in dissemination and, on a broader level, in the way it does not fulfill some of the desirable functions of traditional cooking methods.

The most graphic presentation of the unintended negative effects of projects is not in the area of "appropriate technology," however, but is described by Helen Henderson in her discussion of an AID-financed project in Upper Volta. At issue is the impact on rural women of mobilizing them and raising their expectations that some work will be done to help them improve their lives, and then not doing anything for them at all. USAID probably had excellent reasons for cancelling the second phase of this project, as they had for cancelling the funding of the five villages of Dilly in Mme. Kantara's project. But when a donor organization—or the national government itself—does this, the untold negative effects on the lives of women who already face very unfavorable conditions are immense. At the very least, they cannot be expected to try again to make the extra effort to bring about changes in their life and they will have lost confidence in anything new.

Of course the unintended impacts of projects can be good as well as bad and frequently there are such positive repercussions. Mme. Ki-Zerbo's discussion of the way in which work on the construction of stoves improves relations between men and women is an example. As a second example, Mme. Djire implies in her paper (but also explained in her verbal presentation at Bamako) a whole chain of interrelated positive and mutu-

CONCLUSION 195

ally supportive reactions of women participating in the functional literacy program which were not initially foreseen when the project was established.

Unfortunately, it sometimes seems to observers in the Sahel that the impact of mistakes in development projects outweighs the positive effects which other projects have had. In any case, the cost of failure is very high where people are so poor and resources so scarce.

Solutions and Perspectives

In the late spring of 1985, a little rain had fallen throughout much of the Sahel. Women were busy at the occupations traditionally associated with that time of year. Those engaged in agriculture, with the assistance of male relatives, had prepared the land and planted millet (or corn, or peanuts, etc.) in their own and the communal family fields. They also had carried out all the other tasks of rural women—fetching and carrying water and firewood, gathering and processing wild plants, grinding grain and preparing meals, feeding guinea hens or small ruminants, keeping the home environment in order and the family clothes clean and caring for their children. The normal burden they still bear seems heavy to an outsider, just as their day seems long—often four in the morning until long after sundown in the evening.

What is the prognosis for changing this situation, for lessening the amount of time and physical effort demanded of them and making what they do more productive and more beneficial to them? The short-run outlook is not very favorable. Not much can or will be done immediately for most of the rural women in the Sahel. Although new programs are being introduced, progress is slow.

The papers in this book illustrate the three types of factors which dictate the success or failure of such development efforts in the Sahel. The first are uncontrollable elements which are often unpredictable as well. Natural disasters such as drought fall into this category, as do political factors such as the orientation of the government or the policies pursued by international agencies and other foreign donors. None of these things is in the hands of those Sahelians who are working to bring about development. They can only try to structure their projects so that they will not be destroyed by such unforeseen events, outside intervention, or even the cessation of funding.

The second category includes factors whose influence on project performance is knowable and even measurable. Here are many of the major project elements discussed in the papers—wells and the provision of water, selected seeds and adequate natural and chemical fertilizers, animal-drawn carts, improved roads, adequate market outlets, good credit schemes reaching even poor women farmers, incentives for increased food crop production, training programs geared to general skills capacitation or to specific local needs in agriculture including rain fed crops, vegetable crops and quick growing trees for firewood or animal husbandry for large and small animals, and well-trained and sympathetic *animatrices* and extension agents in sufficient numbers to reach even remote small villages. All of these factors can help bring about development and can improve the lot of rural women. But the constraints are money and trained manpower, both very hard to come by. If more were available, more would be achieved; it is that simple.

Finally, there are factors which are controllable but whose impact is not completely understood. All of the issues debated above fall into this group. The role of cooperatives, the value of functional literacy programs, the degree of specific focus required to make a rural project effective, the importance of improved domestic technologies and the degree to which they are appropriately adapted to local conditions are all examples of things we still do not know enough about. The planner can only tread carefully, trying to use the experience of other programs to avoid problems which have arisen elsewhere. The planner also needs to recognize more clearly the trade-offs between different choices.

Rural development is a complex and delicate process in the Sahel. It is too easy to be discouraged by the slowness of change and by all the uncertainties facing those responsible for programs. But rural Sahelian women, indeed all rural inhabitants, desperately need assistance. In 1985 the Ethiopian drought led to massive starvation and death. At least some part of this disaster could have been avoided by more and better programs for the development of rural areas. The Sahel also is considered by donor countries, the United Nations, and the World Bank as a crisis area threatened, like Ethiopia, by a breakdown of agriculture. Something must be done even if there is no completely satisfactory solution within our grasp.

Appendix

Facts on the Sahel

The word "Sahel" is not found in old American dictionaries although it is in French dictionaries where it is cited as: "Arabic word meaning shore, regions close to the coasts in Algeria and in Tunisia. The word is also used to describe the zone bordering the Sahara in the south" (Editor's translation, *Petit Larousse* 1965, 1667). The latter sentence explains the current common meaning of "Sahel" in English as well as French.

In this book, the Sahel specifically refers to countries which, following the disastrous drought of the early 1970s, joined together in 1973 to form the Comité International de la Lutte contre la Sécheresse au Sahel (CILSS). This organization coordinates and channels aid and research efforts by many donors to the region, although all member countries have continued to have their own bilateral agreements with donor nations and agencies. The nine countries which joined CILSS include: Cape Verde, Chad, Gambia, Guinea-Bissau, Mali, Mauritania, Niger, Senegal, and Upper Volta (Burkina Faso).

Figure A-1

Facts on the Sahel

Countries of the Sahel*	Population (millions)	Area (thousands of sq. kilometers)	GNP per capita (dollars)	Life expectancy at birth	Average index of food production per capita (1969–1971 = 100)	Cereal imports (thousands of metric tons)	Males in primary school (as percentage of age group)	Females in primary school (as percentage of age group)	Urban population (as percentage of total population)
Cape Verde†	.3	4.0	300	60					
Chad	4.6	1,284	80	44	95	57	51	19	19
Gambia‡	.6	11.3	330	m 32 f 34					
Guinea-Bissau‡	.8	26.4	170	35					
Mali	7.1	1,240	180	45	83	143	35	20	19
Mauritania	1.6	1,031	470	45	73	219	43	23	3
Niger	5.9	1,267	310	45	88	120	29	17	6
Senegal	6.0	196	490	44	93	492	58	38	11
Upper Volta (Burkina-Faso)	6.5	274	210	44	95	98	26	15	5

*Source: World Bank 1984, pp. 218, 228, 260, 266.
†Source: Department of State 1984.
‡Source: Department of State 1982.

Bibliography

Acharya, Meena. 1982. *Time Use Data and the Living Standards Measurement Study.* LSMS Working Paper No. 18, Washington: The World Bank.

Agency for International Development, U.S. 1975. *Development Assistance Program 1976–1980—Central West Africa.* Washington, D.C.: Department of State (November).

———. 1976. *Opportunity for Self-Reliance: An Overview of the Sahel Development Potential.* Washington: USAID.

———. 1978. *Report on Women in Development.* Washington: Office of Women in Development, USAID.

———. 1979. *Sahelian Africa: Program Summary 1978.* Washington: Department of State, USAID.

———. 1981. *Selected Statistical Data by Sex: Africa: Mali.* Washington: Office of Women in Development, USAID.

———. 1982. *Women in Development: Proceedings and Papers of the International Conference on Women and Food at the University of Arizona, 1978.* Vols. I and II. Washington: Office of Women in Development, USAID.

Ahmed, Manzoor, and Philip H. Coombs, eds. 1975. *Education for Rural Development: Case Studies for Planners.* New York: Praeger.

American Friends Service Committee. 1982. "Women and Development Programs." Philadelphia: AFSC.

―――. 1981. "Women in Development Programs," in *Annual Report*. Philadelphia: AFSC.

Ames, D. W. 1953. *Plural Marriages among the Wolof in the Gambia*. Evanston: Northwestern University. Available as microfilm from Xerox, University Microfilm.

Amin, Samir. 1980. *Accumulation on a World Scale: A Critique of the Theory of Underdevelopment*. New York: Monthly Review Press.

Approvecho Institute. 1981. *Cookstove News* 1, no. 3 (November).

"Aspecten van internationale samenwerking." 1983. *Voorlichtingsdienst Ontwikkelingssamenwerking* 18, no. 3. Den Haag: Ministerie van Buitenlandse Zaken.

Banerjee, Jayanti. 1981. *Cooking With Firewood—The Burning Issue*. New Delhi: Institute of Social Studies Trust (February).

Barnum, H., L. Squire. 1979. *A Model of an Agricultural Household: Theory and Evidence*. World Bank Occasional Papers #27. Baltimore, Md.: The World Bank.

Barzin-Tardieu, Danielle. 1975. *Les Femmes du Mali*. Ottawa.

Belloncle, G. 1980. *Femmes et developpment en Afrique sahelienne. L'expérience nigérienne d'animation féminine* (1966–1976). Paris: Les Editions Ouvrières.

―――. 1975. *Problems posés par la promotion de la femme rurale en Afrique de l'Ouest: les leçons de l'expérience Nigérienne d'animation féminine*. Paris: Institut de Recherches et d'Applications des Methodes de Developpement.

Benoit-Cattin. 1977. "La Mécanisation des Exploitations Agricoles au Sénégal. Le Cas des Unités Expérimentales du Sine-Saloum." Bambey: CNRA.

Blumberg, Rae Lasser. 1981. "Females, Farming and Food: Rural Development and Women's Participation in Agricultural Production Systems," in Barbara Lewis ed. *Invisible Farmers: Women and The Crisis in Agriculture*. Washington: Office of Women in Development, AID, 23–102.

―――. 1981. *Women at Work in Mali: The Case of the Markala Cooperative*. Working Papers No. 50. Boston: African Studies Center, Boston University.

Boserup, Esther, 1980. "The Position of Women in Economic Production and in the Household with Special Reference to Africa," in Clio Presvelou and Saskia Spijkers-Zwart. eds. *The Household, Women, and Agricultural Development*. Wageningen: H. Veenman en Zonen, 11–16.

―――. 1970. *Women's Role in Economic Development*. New York: St. Martin's Press, 1970.

Bryant, Coralie, and Louise G. White. 1982. *Managing Development in the Third World*. Boulder, Col.: Westview.

Cardoso, Fernando Enrique and Enzo Faletto. 1979. *Dependency and Development in Latin America*. Berkeley: University of California Press.

Carr, Marilyn. 1978. *Appropriate Technology for African Women.* The African Training and Research Centre. Economic Commission for Africa. Addis Ababa: United Nations.

———. 1984. *Blacksmith, Baker, Roofing-Sheet Maker . . . Employment for Rural Women in Developing Countries.* London: Intermediate Technology Publications.

Caughman, Susan. 1983. *Women and Development in Mali, An Annotated Bibliography.* Addis Ababa: ATRCW/UNEZA.

Cernea, Michael. 1979. *Macrosocial Change, Feminization of Agriculture and Peasant Women's Threefold Economic Role.* Washington: The World Bank. Agriculture and Rural Development Department.

Chambers, Robert. 1983. *Rural Development; Putting the Last First.* London: Longman.

Charleton, Sue Ellen M. 1984. *Women in Third World Development.* Boulder, Col.: Westview.

Cheema, G. Shabbir, and Dennis A. Rondinelli, eds. 1983. *Decentralization and Development; Policy Implementation in Developing Countries.* London: Sage Publications.

CILCA. 1983a. *La Formation et l'Animation des Femmes Rurales. Bamako, Mali, 7–9 June 1983.* Bamako: CILCA, B. P. 2652.

———. 1983b. *Learning from Experience.* Report of Harare Workshop and Board Meeting, August 1983. Waltham, Mass.: CILCA, Brandeis University.

Cisse, Aly. 1983. "Compte rendue des Ve Journées intérnationales du Comité Intérnational de Liaison du Corps pour L'Alimentation sur la formation et l'animation des femmes rurales, *7–9 juin.*" In CILCA. 1983a.

Clarke, G. D. et al. 1980. *Local Level Planning and Rural Development: Alternative Strategies.* For the United Nations Asian and Pacific Development Institute. New Delhi: Concept Publishing Co.

Cleve, John. 1974. *African Farmers: Labor Use in the Development of Smallholder Agriculture.* New York: Praeger.

Cloud, Kathleen. 1978. "Sex Roles in Food Production and Distribution Systems in the Sahel." *Women in Development; Proceedings and Papers of the International Conference on Women and Food at the University of Arizona, Tucson.* Washington: Office of Women in Development, USAID.

———. 1985. Women Farmers in AID Agricultural Projects (forthcoming).

———. 1985. Women's Productivity in Agricultural Systems; Considerations for Project Design (forthcoming).

Consortium for International Development. 1980. Final Report of Upper Volta Village Livestock Project. Contract AID/afr-C-1338. Tucson, Arizona.

Coombs, Philip. 1974. *Attacking Rural Poverty: How Non-Formal Education Can Help.* Baltimore, Md.: Johns Hopkins University Press.

Creevey, Lucy. 1980. "The Food Corps in the Sahel," *The Bellagio Report*. Papers and Proceedings from CILCA's First International Workshop July, 1979. Waltham, Mass.: CILCA, Brandeis University.

———. 1980. "Religion and Modernization in Senegal, 1960–1975." In John L. Esposito, ed. *Islam and Development; Religion and Sociopolitical Change*. Syracuse: Syracuse University Press, 207–221.

Creevey (Behrman), Lucy. 1970. *Muslim Brotherhoods and Politics in Senegal*. Cambridge, Mass.: Harvard University Press.

Coulibaly, M. and Mme. 1981. "Evaluation of the Project FEDEV." Philadelphia: American Friends Service Committee.

Darrow, Ken. 1979. "Foreword." In Ianto Evans, *Lorena Owner-Built Stoves*. Stanford, Calif.: Volunteers in Asia, January.

Department of State. 1982. *Background Notes*. Washington, D.C.: U.S. Government Printing Office (November).

———. 1984. *Background Notes*. Washington, D.C.: U.S. Government Printing Office (June).

Dey, J. 1981. "Gambian Women: Unequal Partners in Rice Development Projects?" *The Journal of Development Studies* 17, no. 3 (April): 109–22.

Diaroumeye, Fatoumata A. 1983. "Note de reflexion sur la formation et l'animation des femmes rurales en afrique." CILCA 1983a.

Dicko, Massaran Konate. 1983. "Discours d'Ouverture." See CILCA 1983a.

Dixon, Ruth. 1983. "Women in Agriculture: Counting the Labour Force in Developing Countries." *Population and Development Review* 8, no. 3: 539–66.

Dumont, Bernard. 1973. *Functional Literacy in Mali*. Paris, UNESCO.

Dupire, Marguarite. 1963. "The Position of Women in a Pastoral Society." In D. Paulme, ed. *Women of Tropical Africa*. London. Routledge and Kegan Paul.

El-Sanabary, Nagat M. 1983. *Women and Work in the Third World: The Impact of Industrialization and Global Economic Interdependence*. Berkeley: Center for the Study, Education and Advancement of Women, University of California.

Ernst, K., 1976. *Tradition and Progress in the African Village; The Non-Capitalist Transformation of Rural Communities in Mali*. London.

Esman, Milton J., and Norman T. Uphoff. 1984. *Local Organizations; Intermediaries in Rural Development*. Ithaca: Cornell University Press.

Foltz, William. 1965. *From French West Africa to the Mali Federation*. New Haven: Yale University Press.

Fortman, Louise. 1978. *Women and Tanzanian Agricultural Development*. Economic Research Bureau, Paper 77.4. University of Dar es Salaam.

Friedmann, John, and Clyde Weaver. 1979. *Territory and Function; The Evolution of Regional Planning*. Berkeley: University of California Press.

Galjart, Benno, and Dieke Buijs, eds. 1982. *The Participation of the Poor in Development*. London: Institue of Cultural and Social Studies. University of Leiden.

Ghai, Dharam et al. 1979. *Agrarian Systems and Rural Development*. New York: Holmes & Meier Publishers, Inc.

Goody, J. 1977. *Production and Reproduction. A Comparative Study of the Domestic Domain*. Cambridge: Cambridge University Press.

—— and J. Buckley. 1973. "Inheritance and Woman's Labour in Africa." *Africa* 43, no. 2.

Gordon, David C. 1968. *Women of Algeria: An Essay on Change*. Cambridge: Harvard University Press.

Gould, Helen, and Stephen Joseph. "Designing Stoves for Third World Countries." Intermediate Technology Development Group, London (mimeograph).

Gow, David et al. 1979. *Local Organizations and Rural Development; A Comparative Reappraisal*. 2 vols. Washington: Development Alternatives, Inc.

Griffin, Keith. 1976. *Land Concentration and Rural Poverty*. New York: Holmes & Meier.

Hafkin, N., and Edna Bay. 1976. *Women in Africa: Studies in Social and Economic Change*. Stanford, Calif.: Stanford University Press.

Hammond, P. B. 1966. *Yatenga; Technology in the Culture of a West African Kingdom*. New York: Free Press.

Hanger, Jen, and Jon Moris. 1973. "Women and the Household Economy." In *An Irrigated Rice Settlement in Kenya*. Robert Chambers and Jon Moris, eds. Munich: Weltforum Verlag.

Hemmings, G. N.D. "Draft Paper: Case Studies: Department of the East, Upper Volta." Funded by Women in Development Office. Washington, D. C.

Henderson, H. 1980. "The Role of Women in Livestock Production; Some Preliminary Findings" in Richard Vengroff ed. *Upper Volta: Environmental Uncertainty and Livestock Production*. Lubbock, Tex.: International Center for the Study of Arid and Semi-Arid Lands.

Hill, P. 1972. *Rural Hausa, a Village and a Setting*. Cambridge: Cambridge University Press.

Hirschman, Albert V. 1968. *The Strategy of Economic Development*. New Haven: Yale University Press.

Hopkins, Nicholas S. 1972. *Popular Government in an African Town*. Chicago: University of Chicago Press.

Hobsbawm, Eric. 1964. *Pre-Capitalist Economic Formations*. London: Lawrence Wishart.

Intech, Inc. 1977. *Nutrition Strategy in the Sahel; Final Report*. Washington: Contract AID/TA-C-1214, W010.

International Course for Development Oriented Research in Agriculture. 1982. "Le système de production en pays Serer au Sénégal." Wageningen.

Jayawera, Sivarna et al. 1979. *Status of Women: Sri Lanka*. Colombo: University of Colombo.

Jeffalyn Johnson and Associates. 1980. *African Women in Development; Final Report*. Washington: Office of Regional African Affairs, USAID.

Jequier, Nicholas. 1981. "Appropriate Technology Needs Political 'Push'," *World Health Forum*. Vol. 2., No. 4.

Jeune Afrique. 1973. *The Atlas of Africa*. 1973. New York: The Free Press.

Johnston, Bruce F., and Peter Kilby. 1975. *Agriculture and Structural Transformation*. New York: Oxford University Press.

——— and William C. Clarke. 1982. *Redesigning Rural Development: A Strategic Perspective*. Baltimore: John Hopkins Press.

Jones, Christine. 1982. "Women's Labor Allocation and Irrigated Rice Production in North Cameroon." Jakarta, Indonesia: Paper Prepared for the International Association of Agricultural Economists.

Jones, William I. 1976. *Planning and Economic Policy; Socialist Mali and Her Neighbors*. Washington.

Jouvre, Edmond. 1974. *La Republique du Mali*. Paris.

Ki-Zerbo, Jacqueline, and Guido de Lepeleire. 1979. "L'Amélioration des foyers pour l'utilization domestique du bois de feu: ses possibilités et son impact au Sahel." CILSS et Club de Sahel. (May).

Lahuec, Jean Paul. 1970. "Une communauté évoluée mossi: zaongho (Haute Volta)." *Etudes Rurales* 37–39, pp. 151–172.

Lele, Uma. 1981. "Cooperatives and the Poor: A Comparative Perspective." *World Development* 9(1): 55–72.

———. 1975, *The Design of Rural Development. Lessons from Africa*. Baltimore: the Johns Hopkins University Press.

Leonard, David, and Dale Rogers Marshall. 1982. *Institutions of Rural Development for the Poor*. Berkeley: Institute of International Studies.

Lewis, Barbara. ed. 1982. *Invisible Farmers: Women and the Crisis in Agriculture*. Washington: Office of Women in Development, USAID.

Lewis, John Van D. 1979. "Descendants and Crops; Two Poles of Production in a Malian Peasant Village." New Haven, Ct. Ph.D. Thesis in Anthropology. Yale University.

———. 1981. Domestic Labour Intensity and the Incorporation of Malian Peasant Farmers into Localized Descent Groups." *American Ethnologist* 8(2): 52–73.

Lewis, W. A. 1955. *The Theory of Economic Growth*. London: Allen and Unwen.

Lofchie, Michael, and Stephen K. Commins. 1982. "Food Deficits and Africultural Policies in Tropical Africa." *The Journal of Modern African Studies* 20, no. 1 (March): 1–26.

Ly, Fakaney. 1975. "Mali: Educational Options in a Poor Country," in Ahmed and Coombs, eds. *Education for Rural Development: Case Studies for Planners.* New York: Praeger. 217–247.

McNeil, Leslie Sophia. 1979. "Women of Mali; A Study of Sexual Stratification." Cambridge, Mass. B.A. Thesis, Harvard University.

McRobie, George. 1981. *Small is Possible.* London: Jonathan Cape.

McSweeney, Brenda Gael. 1979. "Collection and Analysis of Data on Rural Women's Time Use." *Studies in Family Planning* 10, no. 11–12 (Nov.–Dec.): 379–383.

Matlock, W. Gerald, and E. Wendell Cockrum. 1976. "Agricultural Production Systems in the Sahel," in *The Politics of Natural Disaster.* New York: Praeger.

Megahed, Horeya T. 1970. *Socialism and Nation-Building in Africa; The Case of Mali 1960–1968.* New York: Praeger.

Meillassoux, Claude. 1971. "Introduction." C. Meillassoux ed., *The Development of Indigenous Trade and Markets in West Africa.* Oxford: Oxford University Press.

Mellor, John W. 1966. *The Economics of Agricultural Development.* Ithaca: Cornell University Press.

Ministerie van Buitenlandse Zaken (MIN). 1983. "Vrouw en voeding in outwikkalings landin, Aspecten van internationale samenwerking," *Voorlichtingsdienst Ontwikkelings. Samenwerking* 16(3).

Monson, Jamie, and Marian Kalb, eds. 1985. *Women as Food Producers in Developing Countries.* Los Angeles: African Studies Center.

Monteil, V. 1967. "The Wolof Kingdom of Kayor." D. Forde and P. M. Kaberry, eds., *West African Kingdoms in the Nineteenth Century.* London: Oxford University Press.

Morgenthau, Ruth Schachter. 1985. Fighting Hunger—A Village in Mali." *Boston Review* (February).

———. 1964. *Political Parties in French-Speaking West Africa.* Oxford: Clarendon Press.

——— and Lucy Creevey. 1984. "French-speaking West Africa." Michael Crowder ed. *Modern Political History of Africa.* London: Cambridge University Press.

Moton, G. 1974. "La Mise en valeur des vallées des Volta Blanche et Rouge en Haute Volta." *Actual Development* 4 (November): 44–50.

Nelson, Nicci. ed. 1981. *African Women in the Development Process.* London: Frank Cass.

Niang and Richard. 1978. "Evolution des principaux facteurs d'intensification dans l'Unité Expérimentale de Thysse-Kaymor, Sonkarong. Kaolack: ISRA.

Nicolaisen, Johannes. 1963. *Ecology and Culture of the Pastoral Tuareg.* National Museum of Copenhagen.

Nurkse, R. 1953. *Problems of Capital Formation in Under-Developed Countries.* London: Oxford University Press.

Obbo, C. 1980. *African Women: Their Struggle for Economic Independence.* London: Zed Press.

van den Oever-Pereira, Petronella. 1979. "Training Women in Rural Africa: A Sahelian Case Study." Ithaca: Master's Thesis, Cornell University.

Overholt, Catherine, and Mary Anderson, Kathleen Cloud, James Austin eds. 1985. *Gender Roles in Development Projects; A Case Book.* West Hartford, Ct.: Kamarian Press.

Pala, Achola. 1976. *African Women in Rural Development: Research Trends and Priorities.* Overseas Liaison Committee Paper No. 12. Washington: American Council on Education.

Paques, V. 1954. *Les Bambara.* Paris: Presses Universitaires de France.

Paulme, Denise. 1963. *Women of Tropical Africa.* London: Routledge and Kegan Paul.

Petit Larousse. 1965. Paris: Auge, Gillon, Hellier-Larousse et Cie.

Pollet, E., and G. Winter. 1968. "L'organisation sociale du travail agricole des Soninke (Dyahunu, Mali)." *Cahiers d'Etudes Africaines* 32, 8(4): 509–534.

Presvelou, Clio, and Saskia Spijkers-Zwart. eds. 1980. *The Household, Women and Agricultural Development.* Proceedings of a Symposium organized by the Department of Home Economics, Agricultural University of Wageningen. Wageningen: H. Veenman en Zonen.

"Project Institut Polytechnique Rurale de Katibougou." 1982. Bamako, Mali: Project Proposal for CILCA.

Raynaut, C. 1977. "Aspects socio-économiques de la préparation et de la circulation de la nourriture dans un village hausa (Niger)." *Cahiers d'Etudes Africaines* 68, 17(4): 569–97.

Republique du Mali. 1981. *Recensement Général de la Population, Décembre 1976.* Bureau Central du Recensement. Bamako.

Riesman, P. 1977. *Freedom in Fulani Social Life.* Chicago: University of Chicago Press.

Rihani, May. 1978. *Development as if Women Mattered; An Annotated Bibliography with a Third World Focus.* Washington: Overseas Development Council.

Robinet, A. H. 1967. "La chevre rousse de Maradi, son exploitation et sa place dans l'économie et l'élévage de la République du Niger." *Rev. Elev. Med. Vet. Pays. Trop.* 20, 129.86.

Rondinelli, Dennis, and Kenneth Ruddle. 1976. *Urban Function in Rural Development; an Analysis of Integrated Spatial Development Strategy.* Washington: Office of Urban Development, USAID.

Rothko Colloquium. 1979. *Towards a New Strategy for Development.* New York: Pergamon Press.

Rupp, Marieanne. 1976. "Report of the Sociological Study Conducted in the Districts of Tanout, Dakoro, Agadez from March 30 to April 30, 1976." Washington: USAID, unpublished, 44 pages.

Sacks, Carol. 1975. "Engels Revisited" in *Women, Culture and Society*. Palo Alto: Stanford University Press.

Sanday, Peggy. 1975. "Female Status in the Public Domain" in *Women, Culture and Society*. Palo Alto: Stanford University Press.

Schoefp, Brooke G. 1979. "Enquête sur la formation des moniteurs agricoles du Mali; rapport de la phase III," unpublished manuscript.

Shaller, Dale V. 1979. "A Sociocultural Assessment of the Lorena Stove and its Diffusion in Highland Guatemela." in Ianto Evans. *Lorena Owner-Built Stoves*. Stanford, Calif.: Volunteers in Asia (January).

Simmons, Emily. 1976. "Calorie and Protein Intakes in Three Villages of Zaria Province, May 1970–July 1971." *Samari Miscellaneous Papers* (Nigeria) 55.

Sjostrom, Margarita, and Rolf Sjostrom. 1982. *How Do You Spell Development? A Study of a Literacy Campaign in Ethiopia*. Uppsala: Scandinavian Institute of African Studies.

Skinner, Elliot. 1964. *The Mossi of Upper Volta*. Stanford, California: Stanford University Press.

Smale, Melinda. 1980. "Women in Mauritania: The Effect of Drought and Migration on their Economic Status and its Implications for Development Programs. Report for AID/WID and USAID Mauritania PASA AG/MAU 300–1–80." Washington: USAID.

Smith, M. G. 1955. *The Economy of Hausa Communities of Zaria*. London: Published by Her Majesty's Stationary Office for the Colonial Office.

Snyder, G. 1965. *One Party Government in Mali*. New Haven: Yale University Press.

Spencer, Dustin. 1976. "African Women in Agricultural Development: A Case Study in Sierra Leone." Overseas Liaison Committee Paper No. 9. Washington: American Council on Education.

——— and Derek Byerlee. 1976. "Technical Change, Labor Use and Small Farmer Development: Evidence from Sierra Leone." *American Journal of Agricultural Economics*. 874–880.

Stohr, Walter B., and D. R. Fraser Taylor. 1981. *Development from Above or Below? The Dialectics of Regional Planning in Developing Countries*. New York: John Wiley.

Streeten, Paul. 1981. *First Things First: Meeting Basic Human Needs in Developing Countries*. London: Oxford University Press.

Streeten, Paul, 1979. "Development Ideas in Historical Perspective." In *Toward a New Strategy for Development;* A Rothko Chapel Colloquium. New York: Pergamon Press.

Taylor, Ellen. 1981. "Women Paraprofessionals in Upper Volta's Rural Development." Special Services on Paraprofessionals. Ithaca: Cornell University, Center for International Studies.

United Nations. 1974a. "The Data Base for Discussion of the Interrelations Between Integration of Women in Development, Their Situation and Population Factors in Africa." Addis Ababa: Economic Commission for Africa.

———. 1974b. *The Role of Women in Population Dynamics Related to Food and Agriculture and Rural Development in Africa*. Economic Commission for Africa/Food and Agriculture Organization, Women's Program Unit. Addis Ababa: Economic Commission for Africa.

———. 1974C. *Plan of Action for the Integration of Women in Development in Africa*. Addis Ababa: United Nations, Economic Commission for Africa.

———. 1975. "The Situation of Women in the Light of Contemporary Time-Budget Research." Report submitted to the World Conference of the International Women's Year. Mexico.

Venema, Bernhard. 1982. "Les Conséquences de l'introduction d'une culture de rente et d'une culture attelée sur la position de la femme Wolof à Saloum." *Revue du Tiers Monde* 23, no. 91: 602–615.

———. 1981. "L'introduction de la traction bovine chez les Wolof du Saloum (Sénégal)." *Etudes Rurales* 84: 87–99.

———. 1980. "Male and Female Farming Systems and Agricultural Intensification in West Africa: The Case of the Wolof, Senegal," C. Presvelou and S. Spijkers-Zwart, eds. *The Household, Women and Agricultural Development*. Wageningen: H. Veenman en Zonen, 27–34.

———. 1978. "The Wolof of Saloum: Social Structure and Rural Development in Senegal." Wageningen: PUDOC.

Vengroff, R. 1980. *Upper Volta: Environmental Uncertainty and Livestock Production*. Lubbock, Tex.: International Center for the Study of Arid and Semi-Arid Lands.

Volunteers in Technical Assistance (VITA). 1980. *Wood Conserving Cook Stoves: A Design Guide*. Mt. Ranier, Md.: VITA, Inc.

Weil, Peter. 1976. "The Staff of Life: Food and Female Fertility in a West African Society." *Africa* 46, no. 2:182–195.

Wilson, Geoffrey, and Monica. 1945. *The Analysis of Social Change; Based on Observations in Central Africa*. Cambridge: Cambridge University Press.

Women, Land and Food Production. 1979. ISIS *International Bulletin* (Spring).

World Bank. 1976. "Rural Development: A Sector Policy Paper." Washington, D.C.: The World Bank.

———. 1985. *World Development Report 1984*. London: Oxford University Press.

Index

American Friends Service Committee, 67, 101.
Animal husbandry, 29, 31–33, 46, 73–74, 118–30, 137–45.
Animation, 5, 13, 92, 97, 115–16, 118–30, 156, 174, 189.
Animation féminine, 40, 45.
Appropriate technology, 1, 10, 75, 77–79, 87, 90, 97, 125, 167–87, 190–94; improved stoves, 10, 97, 169, 174–77, 181–88.

Bamako (Mali), 56, 66, 69, 77, 98, 105, 126, 154, 160, 165.
Bamako Workshop, ix, xii, 4–5, 12, 51, 117, 190–95. *See also* CILCA.
Bambara, 56–60, 66, 69, 73, 76–77, 84–85, 91, 93, 107, 122, 129. *See also* Mali: ethnic groups.
Bobo, 69, 74, 76. *See also* Mali: ethnic groups.
Boserup, Esther, 2, 12, 16, 41, 84, 87–89.
Burkina Faso. *See* Upper Volta.

Cash crops, 1–2, 6–7, 13, 16, 30–31, 38, 40, 56, 74, 84–86.
Catholic Relief Services, 46.
Centres d'Animation Rurales (CAR): in Mali, 67, 71, 75, 101, 108, 110.
CFAR-UNFM. *See* Ouélessébougou.
Chambers, Robert, 2.
CILCA (Comité International de Liaison du Corps pour l'Alimentation), ix-xiii, 4–5, 13, 51, 169.
Cisse, Aly, 65.
Cloud, Kathleen, 16–17, 19, 133, 190.
Comité International de la Lutte contre la Secheresse au Sahel (CILSS), 175, 197.
Cooperatives, 88–89, 99, 149, 159–165, 191; in Mali, 99, 159–65, 191.
Creevey, Lucy, ix, 18, 51, 190.

Daba (hoe), 75, 162.
Development planning (rural development), 2, 5–13, 189–96; "Basic Human Needs," 7–8; "bottom-up versus top-down," 7, 134; "integrated rural development," 8, 191.

209

Diallo, Sacko Coumbo, 65, 99, 159–60, 191.
Dilly (Mali), 123–24, 194.
Djire, Dembele Sata, 65, 99, 153–54, 191.
Dogon, 56, 59. See also Mali: ethnic groups.
Domestic technology. See Appropriate technology.

Education. See Functional literacy.
European Development Fund (also European Fund for Economic Development), 20, 40, 82.

Food crops (also subsistence crops), 1–2, 7, 13, 16, 37, 47, 85.
Fulani, 84, 86, 91, 134–38, 143–47. See also Senegal, Upper Volta.
Functional literacy programs, 68, 97, 99, 111, 153–57.

Haratin, 126. See Mali; ethnic groups.
Haussa (Hausa), 84–86, 91, 93.
Henderson, Helen, 98, 117, 133, 194.

Institute Polytechnique de Katibougou, 56–57.
Institut Sénégalais de Recherche Agricole (ISRA), 81–82.
International Conference on Women and Food (University of Arizona, 1978), 19.

Kantara, Coulibaly Emilie, 65, 98–99, 117, 192–94.
Karite nut (kerite), 28, 75, 161, 172.
Ki-Zerbo, Jacqueline, 100, 167–68, 179, 187, 190, 194.

Koukoundi (Upper Volta), 134-51.
Koulikoro Region (Mali), 66, 69, 99, 106, 156.

Lorena cookstove. See Appropriate technology.
Lusaka Conference 1979, 123.

Mali: agriculture in, 53–65, 67–68; women in agriculture, 17–18, 54–65, 71–75, 123, 161–62; ethnic groups of, 54, 56, 69; government of, 52–53, 108; government leaders: Modibo Keita, 52, 160, Moussa Traore 52–53, 160; history of, 52–65, 68–69, 93; statistics of, xi, 52–54, 66, 68–69, 154, 161, 197–98; regions of, 66, 69. See also Centres d'Animation Rurales, Cooperatives, Functional literacy programs, Opération Riz-Segou, Tons. For women's organizations of Mali, see Union Nationale des Femmes du Mali.
Mandinka, 83, 85–86, 91, 93, 129. See also Senegal.
Maure, 59, 122. See also Mali: ethnic groups.
Minianka, 69, 74, 77. See also Mali: ethnic groups.
Mobilization techniques. See Animation.
Mopti Region (Mali), 66, 69, 99, 156.
Mossi, 56, 107, 141–45, 147–48. See also Mali: ethnic groups, Upper Volta.

National Cooperatives Board (Mali). See Cooperatives.
National Institute of Functional Literacy and Applied Linguistics (INAFLA), 154–55.

Office du Niger (Niger Bureau), 67–68, 70.

INDEX

Opération Riz-Segou (Segou Rice Project), 40, 67, 70, 75.
Ouélessébougou (Mali), 98, 105–16; UNFM training center at, 106–107.

Pastoralists (nomads), 31–35, 56, 122.
Peulh, 54, 59, 69, 74, 122, 126. *See also* Mali: ethnic groups.
Pro-de-so (Stock Farming Project in Western Sahel Region, Mali), 118–31.
Pughtiema, 141–42, 148.

Regional Seminar on Women in Development (Addis Ababa 1974), 44.
Riimaaybe, 138–140, 143–145, 147. See also Upper Volta.
Rupp, Marianne, 33–34, 123.
Rural women: access to land, 42–44, 48, 62–63; inequality of treatment of, 1–2, 10–11, 16, 41–44, 47, 57–58, 62–64, 76, 87–93, 96–97, 190; programs for, 2, 10–11, 16, 40–44, 47, 65, 96–103, 133–35, 151, 195–96.

Sahel, 4, 7, 12, 16, 18, 20, 38–39, 43, 46–47, 96–97, 100–102, 145, 159, 168–69, 177, 179, 189, 195–96; climate and geography, 20, 180; definition of, 20, 197–98; drought in, 19, 38–39, 53, 56, 118; food consumption in, 22–23, 24; food production in, 21–39; women in agriculture in, 19, 26–39.
Sahel Development Program, 39–40.
Samogo, 107. *See also* Mali: ethnic groups.
Sarakolle, 56, 85, 93. *See also* Mali: ethnic groups.
Saudi Arabian Development Fund (Saudi Arabian Development Aid), 98, 119.
Sedentary farmers, 25–31, 56, 122, 137–45.
Segou (Mali), 17, 67–79, 155–56; agriculture in, 69, 73–75; role of women in, 71–78.
Senegal, 4, 46, 81–94, 197–98; women in agriculture in, 82–88; women's organization of, 45; women's projects in, 101–102.
Senoufo, 56. *See also* Mali: ethnic groups.
Serer, 85–86, 93. *See also* Senegal.
Songhay, 56, 59. *See also* Mali: ethnic groups.
Soninke, 122, 126. *See also* Mali: ethnic groups.
Sonono-Bozo, 56, 69, 74, 76. *See also* Mali: ethnic groups.

Thiam, Mariam, x, 17, 57, 67–68, 190.
Tons (farmers' associations, Mali), 109, 129.
Touareg, 59. *See also* Mali: ethnic groups.
Toucouleur, 93. *See also* Senegal.
Traore, Halimatou, 65, 98–99, 105, 117, 192.
Tucker, Jonathan, 100, 167, 194.

UNESCO, 40, 45.
UNFM. See Union Nationale des Femmes du Mali.
UNICEF, 40.
Union Nationale des Femmes du Mali (Union of Malian Women), ix, 5, 45, 67, 97, 105–106, 116, 177. *See also* Ouélessébougou.
United Nations Economic Commission for Africa (ECA), 16, 22, 36, 44.
United Nations Food and Agriculture Organization (FAO), 98, 102, 119.
United Nations Voluntary Fund for Women, 167.
United States Agency for International Development (AID or USAID), 40, 42, 48, 98, 101–103, 124, 131, 133, 194.
Upper Volta (Burkina Faso), 4, 13, 43, 45, 98, 129, 133–52, 177, 185, 187, 194, 197–98; ethnic groups in, 135–45; women's organization of, 45; women's

role in agriculture and animal husbandry, 135–45.
Upper Volta livestock project, 133–51.

Van Dusen Lewis, John, 60–62.
Vegetable gardening (market or truck gardening), 27, 32–33, 42, 46, 56, 73, 97, 123.

Venema, Bernhard, 17, 64, 75–76, 81, 190.
Wolof, 18, 75, 81–93; agriculture of, 81–91. *See also* Senegal.
Women and Food Information Network, 19, 133.
World Bank, 7, 20, 196.
World Food Conference (Rome, 1974), 7.

WOMEN FARMERS IN AFRICA

was composed in 10-point Mergenthaler Linotron 202 Galliard and leaded 2 points
by Coghill Book Typesetting Co.,
with display type in Vero New Antiqua by Rochester Mono/Headliners;
printed sheet-fed offset on 50-pound, acid-free Glatfelter B-31 Natural,
Smyth sewn and bound over binder's boards in Holliston Roxite B,
also adhesive bound with paper covers
by Thomson-Shore, Inc.;
with paper covers printed in 2 colors by Thomson-Shore, Inc.;
and published by

SYRACUSE UNIVERSITY PRESS
SYRACUSE, NEW YORK 13244-5160